HE
PROBLEM

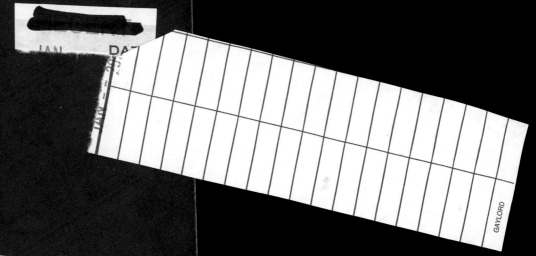

GAYLORD

The MIT Press Essential Knowledge Series

THE MIND–BODY PROBLEM

JONATHAN WESTPHAL

The MIT Press | Cambridge, Massachusetts | London, England

MIT Press books may be purchased at special quantity discounts for business or sales promotional use. For information, please email special_sales @mitpress.mit.edu or write to Special Sales Department, The MIT Press, 1 Rogers Street, Cambridge, MA 02142.

This book was set in Chaparral and DIN by Toppan Best-set Premedia Limited. Printed and bound in the United States of America.

Library of Congress Cataloging-in-Publication Data

Names: Westphal, Jonathan, 1951- author.
Title: The mind-body problem / Jonathan Westphal.
Description: Cambridge, MA : MIT Press, 2016. | Series: The MIT Press essential knowledge series | Includes bibliographical references and index.
Identifiers: LCCN 2016014329 | ISBN 9780262529563 (pbk. : alk. paper)
Subjects: LCSH: Mind and body.
Classification: LCC BF161 .W455 2016 | DDC 128/.2—dc23 LC record available at https://lccn.loc.gov/2016014329

10 9 8 7 6 5 4 3 2 1

CONTENTS

SERIES FOREWORD

The MIT Press Essential Knowledge series offers accessible, concise, beautifully produced pocket-size books on topics of current interest. Written by leading thinkers, the books in this series deliver expert overviews of subjects that range from the cultural and the historical to the scientific and the technical.

In today's era of instant information gratification, we have ready access to opinions, rationalizations, and superficial descriptions. Much harder to come by is the foundational knowledge that informs a principled understanding of the world. Essential Knowledge books fill that need. Synthesizing specialized subject matter for nonspecialists and engaging critical topics through fundamentals, each of these compact volumes offers readers a point of access to complex ideas.

Bruce Tidor
Professor of Biological Engineering and Computer Science
Massachusetts Institute of Technology

PREFACE

The mind–body problem is a *paradox*. A paradox is a group of propositions for each of which we have apparently sound arguments, yet the propositions taken together are inconsistent. We cannot affirm all the propositions in the group, yet we have good reason to believe that they are all true.

My approach is a narrower one than is usually taken to the mind–body problem, but I believe that this approach can help us to think clearly about what is going on with the particular solutions that have been given to the problem in the past. It will help us not to get lost in the metaphysics of things other than the mind and the body. The mind–body problem, in its full generality, which I introduce in chapter 1, is about the mind and the body, not about the self, or consciousness, or the soul, or anything other than the mind and the body. In chapter 5, however, I do consider some important scientific theories of consciousness as examples of scientific treatments of this part of the mind and the mind–body problem. Consciousness, the study of which has recently become important in the cognitive sciences, can reasonably be regarded as a *part* of the mind, though it is not the whole of it, nor is it as Descartes believed the essence of mind.

Among other questions, the mind–body problem can be taken to raise the issue of the physicality of the mind.

It is certainly hard to see how it can be true that the mind, and with it consciousness, is just physical matter. We cannot watch or even imagine the neurons firing and emitting little bursts of mentality or consciousness, like diaphanous fairies flitting around the brain. The brain is obstinately physical, indeed material, weighing in at a hefty three pounds or so, but it does not make sense even to *ask* how much the mind weighs. One can say that the human brain is approximately 2 percent of the weight of the human being. And one can say that it contains one hundred million neurons, not to mention glial cells. But these things are not true of the human mind. We can say that the brain measures roughly five by six by four inches. Nothing remotely like that can be said of the mind. Of course, there are more sophisticated versions of materialism (the thesis that everything that exists is something material) or physicalism (the thesis that everything that exists is something physical), which I discuss in chapter 3, but the same difficulty remains buried in all of them.

Following David Chalmers, philosophers and others have wanted to distinguish the *hard problem* from the *easy problem* of consciousness, or rather the easy *problems*—they are many, according to Chalmers—and quite a lot has been written about this recently.[1] By the easy problems, Chalmers means the problems of describing the physical processes by which we come to have, for example, the consciousness of whiteness, assuming that we can make sense of this phrase. Clearly there *are* many problems of this sort. We have to understand the mechanisms of the eyes, of the ears, of touch, of the nose, and so on. The hard problem is to understand how our experience of whiteness, and with it our consciousness of whiteness, could arise from the purely physical systems operating in the visual cortex. The idea is that we can understand experience by the physical processes that go on when we perceive, but that there are properties of the experiences that cannot be understood in this way. These are the *qualia*, and for Chalmers they are not physical.

Fair enough, but this new twentieth-century "hard problem" is simply a souped-up version of an old problem which, as we shall see in chapter 1, appeared with Descartes and his critics in 1641. The hard problem *is* hard. Why? Because it is the mind–body problem, and that is a hard problem. Chalmers's hard problem of consciousness *is just the mind–body problem* with a new name, complete with a very sharp distinction between the more easily understood physical processes, and consciousness or qualia, or mind. For Descartes, the principal attribute of the mind is consciousness, and so, as his critics pointed out, there is a problem about the relation of the mind to the body. This is all down to the difficulty of asserting that the mind is physical: the mind does not seem to fit into the world of physics and physiology at all. Or so claims dualism. I consider different forms of dualism in chapter 2.

In the mid-seventeenth century, the mind–body problem became a central problem—"the world-knot," as it is often said that Schopenhauer called it—he didn't.[2] What Schopenhauer would perhaps have meant had he used that phrase for the mind–body problem is that the problem signifies that our day-to-day concepts of the mental and the physical and the mind and the body are somehow tangled up in such a way that our entire conception of the world, including the physical world, is called into question by the mind–body problem. The metaphysical problem itself has only gained in significance since philosophers first became aware of it, and at the end of the twentieth century and the beginning of the twenty-first, for better or for worse, it is still alive and well.

My own view, presented in chapters 6 and 7, is friendly toward antimaterialist arguments. It is also friendly to dualism, the view that there are two distinct kinds of things in the world, not just one.

I have been thinking about the mind–body problem since I first encountered it as a student. At that time, the atmosphere in American and English universities was strongly physicalist or materialist in orientation. It is hard to recapture a sense of how strong that atmosphere was, and the way in which students who held other views felt intimidated and sidelined. I imagine that it would be very difficult now to give a historically true account of how these students moved into disciplines outside philosophy, disciplines that for them had more of a feeling for what kind of thing the mind and even the human being might be, and of what the possibilities of philosophy were. Fortunately, philosophy today is becoming more welcoming to the outsider, as it should be, not just because of a corrective politics, but because philosophy itself has realized that it fundamentally mistakes its mission if it is unwelcoming to the intellectual stranger, the outsider.

I was one of those sidelined students. For me at that time, the powerful contemporary materialism was a problem. I started in freshman year with seventeenth-century philosophy, especially Leibniz, under Professor Edwin Curley, and I felt immediately at home there. I was convinced, however, and I still am, that a really determined and sustained analysis of a philosophical view, coupled with a sense of responsibility to the language in which it is expressed, will end up arriving, at the very least, at the problems faced by that view, and at the most, at a definite knowledge about whether the view is true or false. I have read and heard nothing since I was a student that has made me doubt this conviction. Materialism might have been true. Since the 1960s it has suffered something of a decline, and several significant new antimaterialist arguments have appeared in the philosophy of mind. Chapter 4 is about these arguments. It is interesting and surprising that the new developments have taken place just as philosophy and science are becoming friendlier toward one another. It is even more interesting that these arguments have hardly disturbed the default materialist and naturalist convictions of the bulk of scientists and philosophers today. ("Naturalism" is the view that nature is all there is, so that all occurrences are natural occurrences.) Where do these materialist and naturalist convictions come from, if not from reason? Reason can of course take many forms, and one of them *is* the merely pedestrian or labored intellect.

Whatever view we take of the current situation, the time seems favorable to a study of the *whole* problem. This means more than fussing with debates about the details of the theories that have been offered for the solution of the problem. The problem is not about debates, and philosophy generally is not debate. What we need is an *understanding* of the structure of the problem, and of its origins in the concepts of the mind, the body, the physical, and the mental.

Why another book on the mind–body problem? Why now? The answer is that this is not *another* book on the mind–body problem. There has been no full-length and comprehensive book devoted to the problem for a very long time indeed. I believe that this may be partly to do with the fact that a materialist orientation is still the most natural one for many philosophers. From that point of view, the mind–body problem really is impossibly hard. For me, that is yet another reason not to be a materialist, and instead to look more carefully at forgotten or overlooked views in the history of the subject, as I do in the last two chapters of the present book, arguing for my own view (neutral monism) in chapter 7. I am particularly interested in the fact that neutral monists in the past have not given enough attention, if any, to mind–body interaction, nor have they actually stated a solution to the mind–body problem, contenting themselves instead with enthusiasm about the oneness of things. I have offered some suggestions about how to remedy this, and an analysis of how the neutral monist ought to understand mind–body interaction.

THE MIND–BODY PROBLEM: BACKGROUND AND HISTORY

What Is the Mind–Body Problem?

From a logical point of view, the mind–body problem is easy to understand, and it can be expressed clearly, in just four propositions or statements. The following formulation is one I have adapted from Keith Campbell.[1]

(1) The mind is a nonphysical thing.
(2) The body is a physical thing.
(3) The mind and the body interact.
(4) Physical and nonphysical things cannot interact.

It is very hard to deny any of these four propositions. But they cannot consistently be held to be true together. At least one of them must be false, and the attempt to show the exact way in which this plays out is the work of developing a solution to the mind–body problem.

As formulated above in (1)–(4), the mind–body problem is an entirely logical problem. The four propositions simply *cannot* consistently be maintained together; nothing can change that. There really is a contradiction to be derived from them, and the problem is the tension *between* the propositions. Of course, it is also possible to maintain vague propositions very similar to all four propositions, but which do not have the hard-and-fast relationships that are suggested by the formal terms in which the inconsistent group is stated.

Figure 1 shows how the four terms "mind," "body," "physical," and "nonphysical" are related in the four propositions in such a way as to produce an inconsistency. The

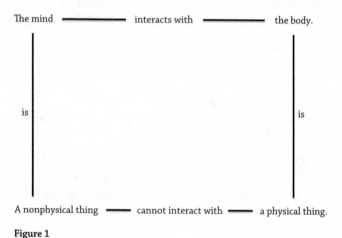

Figure 1

point of putting the problem in this rather formal way is that the four propositions are what philosophers and logicians sometimes call an "inconsistent tetrad." What this phrase means is that of the *four* propositions (the tetrad) any *three* can be true at the same time, but if they are, then the *fourth* is false. Here is the inconsistency. For example, if (1), (2), and (3) are true, then (4) is false. If the mind is a nonphysical thing, and the body is a physical thing, and mind and body interact, then it follows that at least one nonphysical thing and one physical thing *do* in fact interact, and so it is false that physical and nonphysical things cannot interact; and the fourth proposition is false.

Or suppose the last three propositions, (2), (3), and (4), are true. Suppose that physical and nonphysical things cannot interact, that mind and body do interact, and that the body is a physical thing. Then it follows that the mind is not a nonphysical thing, which is to say that the mind is a physical thing. The mind must be a physical thing, because the body is a physical thing, and it interacts with the mind. But physical and nonphysical things, we assumed, cannot interact. So (1) must be false.

It is fun to play around with the four original propositions (1) through (4) in this way, choosing any three, and then deriving the opposite or negation of the fourth. By doing this one can get a good sense of how the mind–body problem is a logical problem. It is a problem that *cannot* be solved, if by solving it one means holding onto all four propositions at once. When one has seen that holding onto all four propositions is not a logical option, one can see clearly that the first and most basic question is *which* one of the four propositions one is going to deny.

We can also describe the mind–body problem in less formal terms. Consider the human body, with everything in it, including internal and external organs and parts, such as the stomach, nerves and brain, arms, legs, eyes, and all the rest. Even with all this equipment, especially the sensory organs, it is interesting and surprising that we can consciously perceive things in the world that are far away from us. For example, I can open my eyes in the morning, and see a nice cup of coffee waiting for me on the bedside table. There it is, a foot away, and I am not touching it, yet somehow it is making itself manifest to me. How does it happen that I *see* it? How does the visual system convey to my awareness or mind the image of the cup of coffee?

The answer is not particularly simple. Very roughly, the physical story is that light enters my eyes from the cup of coffee, and this light impinges on the two retinas at the backs of the eyes. Then, as we have learned from physiological science, the two retinas send electrical signals past the optic chiasm down the optic nerve. These signals are conveyed to the so-called visual cortex at the back of the brain. And then there is a sort of a miracle. The visual cortex becomes active, and I *see* the coffee cup. I am *conscious* of the cup, we might even say, though it is not clear what this means and how it differs from saying that I *see* the cup.

How did the physical state of the brain produce in me the exciting awareness of the presence of the cup of coffee? One minute there are just neurons firing away, and no image of the cup of coffee. The next, there it is; I see the cup of coffee, a foot away. I am not aware of all those neurons firing, and I certainly don't see them. The neurosurgeon is the one who sees *them*. What *I* see is a cup of coffee. How did my neurons contact me or my mind or consciousness, and stamp there the image of the cup of coffee for me? How did the sensation of a cup of coffee arise from the mass of neurons?

It's a mystery.

That mystery is the mind–body problem, or part of it.

We might want to divide the problem up this way. Let us call the "visual experience" the mental state that, in this case, contains the image of the cup of coffee. The mental state is related in some way, still to be determined, to whatever is going on objectively and physically when we see the cup.

In addition to these two sets of events (the mental experience and the physical events that underlie it) there is said to be the "subjective character" of what is seen. This subjective character is something for which some philosophers have adopted the phrase "what it is like," as in the phrase "what it is like to have the experience of seeing a cup." What it is like to have the experience of seeing a cup is to be identified with the *consciousness* of seeing a cup. To be conscious of the cup is for there to be something *it is like* to have that experience. The phrase originates with the philosopher Timothy Sprigge, and was also used later and made popular by Thomas Nagel.[2] It is intended to capture this extra bit of what experience involves. What is it like to see a cup of coffee? Or, in more general terms, what is it like to be a conscious human being?

Compare this with the question of what it is like to be a stone. Well, there is *nothing* it is like to be a stone. So by the criterion of Sprigge and Nagel, the stone has no consciousness.

This "what it is like" has also received from philosophers and others the name *quale* (Latin, singular, pronounced "kwa-lay," to rhyme with "parlay") and *qualia* (Latin, plural, pronounced "kwa-lee-ah," to rhyme with "la-dee-dah").

When I see or otherwise perceive a cup of coffee, I am aware of the quale that attaches to the experience, and which is presumably altogether absent from a video feed carrying the same information. Video displays do not have consciousness. For consciousness to come into the picture, someone has to be *looking* at the video picture.

The idea of the quale is unfortunately not altogether clear. Some philosophers have used the term to refer not to *properties* of experiences, such as the white cup shape that

What is it like to see a cup of coffee? Or, in more general terms, what is it like to be a conscious human being?

I see when I see the cup, as I have outlined their role above, but to the experiences themselves.

There is a genuine ambiguity here, and the two philosophical usages are inconsistent. If qualia are experiences, then they themselves are properties of the subjects who have them, and they are psychological entities. If they are properties *of* experiences, then they are not properties of the subjects who have the experiences; they are independent metaphysical entities, for example *the color white*, and *the shape of the cup*, which somehow turn up in the subjects' minds.

There is also the question whether when I see a white cup I have one quale or many. Do I have one white cup quale, or many smaller white-colored cup qualia that together make up the whole cup image? Neither answer is satisfactory. How many qualia do I have when I look at a speckled hen, to take a famous example? I cannot see how many spots or speckles there are, especially if the hen is running about, with its head bobbing, and there may not even be a definite number. Or how many different qualia do I have when I look at a quality changing smoothly over time, say, the reds across the spectrum? The point is that the same difficulty does not attack concepts like *a color*, or *being colored*, for example in the claim that the bread is brown in color, even though specks of it are white, say, or gray. *Color* and *being colored* are more tolerant concepts than the quale.

I shall use the words "qualia" and "quale" when the context or the author I am discussing calls for them, and "phenomenal property" when the emphasis is more on the quality that is the object of experience. I regret to say that the history of the terminology is sufficiently confused to allow such latitude. The term "qualia" has an interesting and one might say chequered history. In the past the phrases and words "*cogitationes*," "ideas," "experiences," "sense data," "qualities," "perceptions," "sensations," "properties of sensations," "percepts," "raw feels," "nomological danglers," "phenomenal properties," and "qualitative properties" have been used to try to get at something approximately like the same idea. The confused history of the different terminologies is enough to alert the thoughtful student of recent philosophy to the fact that all is not as it should be in the kingdom of the qualia. Why the frequent changes in terminology, and the zigzag of complicated arguments to try to get across what ought to be a fairly straightforward idea, or set of ideas?

There is a well-known story about Herbert Feigl giving a lecture about the mind and the brain at UCLA in 1966, in which he discussed a part of the problem of the relation of mind and brain to which he simply couldn't see the solution, in spite of his materialism, namely the problem of "raw feels." The distinguished philosopher Rudolf Carnap was in the audience, and he announced in the Q&A that he had a solution to the problem of raw feels. Feigl was excited, and asked what it was. "The solution to your problem, Herbert," replied Carnap, "is the α-factor." Feigl got even more excited and wanted to know what the α-factor was, as, in spite of his scientific education, the concept was new to him. "Well, Herbert, you tell me what a raw feel is, and I'll tell you what the α-factor is," Carnap responded. It was a fair point.

The conceptual and linguistic difficulties of describing qualia or phenomenal properties are formidable enough, yet it remains a fact that though a scientist can take a scan of my brain, say, recording my blood flowing or my neurons firing, there is no equivalent scan for my experiences. There do seem to be two different worlds here that are related—but how are they related?

It is also important to be aware that the mind–body problem is about the relationship between the human *mind* and the human physical *body*. (It is also about the relationship between animal minds, if there are any, and animal bodies, but in this book I will restrict discussion to the human case.) The relationship between mind and body exists with or without qualia. If I am in the mental state involved in unconsciously seeing a cup of coffee, or unconsciously thinking about a problem, one might very well wonder how that mental state is related to the physical body, even if no qualia attach to it.

It is easier to see what the problem is if we consider the mind–body problem going in the opposite direction.

Consider the body again. There are my arms, outside the blankets, and I reach out with my right hand and take hold of my cup of coffee, because I want to have a sip of coffee. How did I do that? How did my *mental* desire for a sip of coffee get my *physical* arm to reach out to the cup? Well, we know the answer, at least partly. My muscles moved my arm. But how did my *mental* desire move my physical muscles? Did my mind somehow reach into my arm and move the muscles?

Again, we know the answer, or think we do. Electrical signals from the brain moved the muscles, not mental energy directed at them. Yet the question persists. How did my *mental* desire cause the *physical* electrical signals to start up and then to run down my arms and move the muscles? Again, physiology provides an answer. It was not my mind—"the mental"—that produced the physical electrical signals, but the neurons firing in my brain. All right, but now we get to the nerve of the problem (so to speak). How did my *mental* desire, carrying its associated quale along with it, cause my *physical* neurons to fire? We seem to have some form of telekinesis here, the moving of objects by mental energy alone. If the response is that it is *other* neurons, rather than my mental want, that caused the neurons to fire, then the question has been avoided rather than answered. How does my mental desire cause those *other* neurons to fire?

There might be no qualia associated with this desire. Or there might be an unconscious mental desire. But the question remains of how it causes the neurons to fire and to initiate the moving of my arm. So qualia are part of the mind–body problem, but the problem also involves any relationship between the mind and the body, including unconscious states of mind and physical states.

Matter or the physical can somehow affect the mind; and the mind can somehow move the physical body. These "somehows" are difficult to understand, though, because we cannot see either how there could be aspects of mental experience or qualia floating around amid the neurons, or desires trailing clouds of their attendant qualia, physically digging into the neurons and making them fire.

Descartes and the Discovery of the Problem in 1641

There is a very common view which states that, in the *Meditations on First Philosophy* of 1641, and also in the *Treatise on Man*, written some years earlier, the French philosopher René Descartes discovered, or invented, the problem that today we call the mind–body problem.

Our mind–body problem is not just a difficulty about how the mind and body are related and how they affect one another. It is also a difficulty about how they *can* be related and how they *can* affect one another. Their characteristic properties are very different, like oil and water, which simply won't mix, given what they are.

According to Descartes, matter is essentially *spatial*, and it has the characteristic properties of linear dimension. Things in space have a position, at least, and a height, a depth, and a length, or one or more of these. Areas are two-dimensional, and lines are one-dimensional, but both have a place in space. Objects are three-dimensional, apparently, at least in ordinary experience. Mental entities, on the other hand, do not have these characteristics. We cannot say, of a mind, or any part of it, that it is a two-by-two-by-two-inch cube or a sphere with a two-inch radius, for example, located in a position in space inside the skull. This is not because it has some other shape in space, but because it is not characterized by space at all. What is characteristic of a mind, Descartes claims, is that it is *conscious*, not that it has shape or consists of physical matter. Unlike the brain, which has physical characteristics and occupies space, it does not seem to make sense to attach spatial descriptions to the mind. We can ask, "How much space does the mind occupy?" or "What shape is it?" or "Is it three-dimensional or two-dimensional?" or "Where is it in physical space?" But these questions have no answers, as the questions make no sense.

There is no need to claim, as Descartes did, that *the* essence of the physical is space; we need merely that something's being *in* space is a necessary condition for its being physical at all. It is interesting that this straightforward test of physicality has survived all the philosophical changes of opinion since Descartes, almost unscathed. Even some strange entity in physics that is said not to be in space does not automatically count as a counterexample, for there is nothing to prevent us from saying that, with such entities, physics is dealing with something nonphysical—nonphysical just because it does not have a position in space. And typically, entities that are said not to have a position in space are more the creatures of mathematics than of physics. To drive this point home, we should ponder Noam Chomsky's celebrated view that we do not even know what the mind–body problem is because we do not have a clear concept of the physical or of body: "Lacking a concept of 'matter' or 'body' or 'the physical,' we have no coherent way to formulate issues related to the 'mind–body problem.'"[3] But we *do* have a concept of space laid out before us, and of physics as dealing with whatever it contains. Our bodies are certainly in space, and our minds are not, in the very straightforward sense that the assignation of linear dimensions and locations to them or to their contents and activities is unintelligible.[4]

Such issues aroused considerable interest following the publication of Descartes's *Meditations*, starting with the "Objections" to Descartes. The "Objections" were written by a group of distinguished contemporaries, and in return Descartes wrote his "Replies." The "Objections and

Replies" were included in the first edition of the *Meditations*. Though we do find in the *Meditations* itself the distinction (the "real distinction") between the mind and the body, drawn very sharply by Descartes, in fact he makes no mention of our mind–body problem. Descartes is untroubled by the fact that, as he has described them, mind and matter are very different: one is spatial and the other not, and *therefore one cannot act upon the other*. Something lacking a position in space cannot act upon something in space, say at a point. The problem is simply not there in his text. Descartes himself writes in his Reply to one of the Objections:

> The whole problem contained in such questions
> arises simply from a supposition that is false and
> cannot in any way be proved, namely that, if the soul
> and the body are two substances whose nature is
> different, this prevents them from being able to act
> on each other.[5]

Descartes is surely right about this. The "nature" of a baked Alaska pudding is very different from that of a human being, no doubt about this at all, since one is a pudding and the other is a human being, but the two can "act on each other" without difficulty. The human being can eat the baked Alaska pudding, for example, and the baked Alaska can give the human being a stomachache.

The difficulty, however, is not merely that mind and body are different. It is that they are different in such a way that their interaction is impossible because it involves a contradiction. It is the nature of bodies to be in space, and the nature of minds not to be in space, Descartes claims. For the two to interact, what is not in space must act on what is in space. Action on a body takes place at a position in space, however, where the body is. So mind, or a bit of it, must get up next to the space inhabited by the body. But (to repeat) minds are not in space and nor are they spatially related to it, so they cannot even get near it.

Apparently Descartes did not see *this* problem. It was, however, clearly stated by two of his critics, his correspondent Princess Elisabeth of Bohemia, and his respondent Pierre Gassendi. They pointed out that if the soul is to affect the body, it must make contact with the body, and to do that it must be in space and have extension. In that case the soul is physical, by Descartes's own criterion.

In a letter dated May 1643, Princess Elisabeth wrote to Descartes,

> I beg you to tell me how the human soul can determine the movement of the animal spirits in the body so as to perform voluntary acts—being as it is merely a conscious substance. For the determination of the movement seems always to come about from the moving body's being propelled—to depend on the

kind of impulse it gets from what it sets in motion, or again, on the nature and shape of this latter thing's surface. Now the first two conditions involve contact, and the third involves that the impelling [thing] has extension; but you utterly exclude extension from your notion of soul, and contact seems to me incompatible with a thing's being immaterial.[6]

Propulsion and "the kind of impulse" that set the body in motion require contact, and "the nature and shape" of the surface of the site at which contact is made with the body require extension. We need two further clarifications to grasp this passage. The first is that when Princess Elisabeth and Descartes mention "animal spirits" (the phrase is from Galen) they are writing about something that plays roughly the role of signals in the nerve fibers of modern physiology. For Descartes, the animal spirits were not spirits in the sense of ghostly apparitions, but part of a theory that claimed that muscles were moved by inflation with air, the so-called balloonist theory. The animal spirits were fine streams of air that inflated the muscles. ("Animal" does not mean the beasts here, but is an adjective derived from "anima," the soul.)

The second clarification is that when Princess Elisabeth writes that "you utterly exclude extension from your notion of soul," she is referring to the fact that Descartes defines mind and matter in such a way that the two are mutually exclusive. Mind is consciousness, which has no extension or spatial dimension, and matter is not conscious, since it is completely defined by its spatial dimensions and location. Since mind lacks a location and spatial dimensions, Elisabeth is arguing, it cannot make contact with matter. It cannot possess a contacting surface or an impulse that operates on an extended surface. Here we have the mind–body problem going at full throttle.

Pierre Gassendi was one of the philosophers and scientists who wrote one of the so-called Objections to Descartes's *Meditations*. He puts one of his criticisms this way:

> For how, may I ask, do you think that you, an unextended subject, could receive the semblance or idea of a body that is extended?[7]

By "semblance" Gassendi means something like what we would call an image. It is worth noting that images, in a perfectly precise photographic sense, are carried by light to the eye. The sense is that a photograph of the object can be taken from anywhere between us and the object, or from any other place at which light carries the information of the image.

Descartes himself did not yet have the mind–body *problem*; he had something that amounted to a *solution* to the problem. It was his *critics* who discovered the problem, right in Descartes's *solution* to the problem, although it is also true that it was almost forced on them by Descartes's sharp distinction between mind and body. The distinction involved the defining characteristics or "principal attributes," as he called them, of mind and body, which are consciousness and extension.

Though Descartes was no doubt right that very different kinds of things can interact with one another, he was not right in his account of how such different things as mind and body do in fact interact. His proposal, in *The Passions of the Soul* of 1649, was that they interact through the pineal gland, which is, he writes, "the principal seat of the soul" and is moved this way and that by the soul so as to move the animal spirits or streams of air from the sacs next to it. He had his reasons for choosing this organ, as the pineal gland is small, light, not bilaterally doubled, and centrally located. Still, the whole idea is a nonstarter, because the pineal gland is as physical as any other part of the body. If there is a problem about how the mind can act on the body, the same problem will exist about how the mind can act on the pineal gland, even if there is a good story to tell about the hydraulics of the "pneumatic" (or nervous) system.

We have inherited the sharp distinction between mind and body, though not exactly in Descartes's form, but we have not inherited Descartes's solution to the mind–body problem. So we are left with the problem, minus a solution. We see that the experiences we have, such as experiences

We see that the experiences we have, such as experiences of color, are indeed very different from the electromagnetic radiation that ultimately produces them, or from the activity of the neurons in the brain.

of color, are indeed very different from the electromagnetic radiation that ultimately produces them, or from the activity of the neurons in the brain. We are bound to wonder how the uncolored radiation can produce the color, even if its effects can be followed as far as the neurons in the visual cortex. In other words, we make a sharp distinction between physics and physiology on the one hand, and psychology on the other, without a principled way to connect them. Physics consists of a set of concepts that includes *mass*, *velocity*, *electron*, *wave*, and so on, but does not include the concepts *red*, *yellow*, *black*, *pink*, and the like. Physiology includes the concepts *neuron*, *glial cell*, *visual cortex*, *membrane potential*, and so on, but does not include the concept *red* and all the other color concepts. The color red is something that we see. In the framework of current scientific theory, "red" is a *psychological* term, not a physical one. Then our problem can be very generally described as the difficulty of describing the relationship between the physical and the psychological, since, as Princess Elisabeth and Gassendi realized, they possess no common relating terms.

Was there really no mind–body problem before Descartes and his debate with his critics in 1641? Of course, long before Descartes, philosophers and religious thinkers had spoken about the body and the mind or soul, and their relationship. Plato, for example, wrote a fascinating dialogue, the *Phaedo*, which contains arguments for the survival of the soul after death, and for its immortality. Yet the exact sense in which the soul or mind is able to be "in" the body, and also to leave it, is apparently not something that presented itself to Plato as a problem in its own right. His interest is in the fact *that* the soul survives death, not *how*, or in what sense it can be in the body. The same is true of the religious thinkers. Their concern is for the human being, and perhaps for the welfare of the body, but mainly for the welfare and future of the human soul. They do not formulate a problem with the technical precision that was forced on Princess Elisabeth and Gassendi by Descartes's neatly formulated dualism.

Something important clearly changed in our intellectual orientation during the mid-seventeenth century. Mechanical explanations had become the order of the day, such as Descartes's balloonist explanation of the nervous system, and these explanations left unanswered the question of what should be said about the human mind and human consciousness from the physical and mechanical point of view. What happens, if anything, for example, when we decide to do even such a simple thing as to lift up a cup and take a sip of coffee? The arm moves, but it is difficult to see how the thought or desire could make that happen. It is as though a ghost were to try to lift up a coffee cup. Its ghostly arm would, one supposes, simply pass through the cup without affecting it and without being able to cause it or the physical arm to go up in the air. It would be no less remarkable if merely by thinking about it from a few feet away we could cause an ATM to dispense cash. It is no use insisting that our minds are after all not physically connected to the ATM, and that is why it is impossible to affect the ATM's output—for there is no sense in which they are physically connected to our bodies. Our minds are not *physically* connected to our bodies. How could they be, if they are nonphysical? That is the point whose importance Princess Elisabeth and Gassendi saw more clearly than anyone had before them, including Descartes himself.

DUALIST THEORIES OF MIND AND BODY

Interactionism and Substance Dualism

Mind–body dualism was a popular view until roughly the 1960s, though it is less and less so these days, at least with professional philosophers. They have for the most part thrown in their lot with those scientists who have adopted a materialistic or naturalistic worldview—nature is all there is.

Dualism is the antinaturalist claim that the mind and the body are two separate and very different things. The two sorts are the *nonphysical* and the *physical*. The nonphysical sort of thing, the mind or soul, is not part of nature. "The mind is a nonphysical thing" was our first proposition, and "The body is a physical thing" the next. The essence of dualism is the claim that both these propositions are true, and that the mind is not part of nature.

Dualism is the anti-naturalist claim that the mind and the body are two separate and very different things. The two sorts are the *non-physical* and the *physical*.

In addition, one important form of dualism tells us that mind and body are distinct *things* that can exist independently of one another. Such independently existing things have been called "substances" in the history of philosophy. A substance is an individual thing that can exist by itself, independently of other substances. Accordingly, *substance dualism* is the view that mind and body are distinct in the sense that they can exist independently of each other, or are substances.

Interactionist substance dualism is the view that these two substances or things exist and can interact causally. So, for example, when the body takes in too much beer, the mind becomes confused, and one's mood may change. Here the interactionist substance dualist will say that the physical substance or thing called "the body" is interacting, or certainly seeming to, with the nonphysical substance or thing called "the mind."

The body can exist without the mind, after burial. But what about the other way round? We can imagine the mind existing in the darkness after death, just as it exists in the darkness after bedtime. Just as vividly as we are aware of our mind in the darkness after lights-out, perhaps working on little mathematics problems, or perhaps saying its prayers, or thinking about this or that, so we can imagine activity of these sorts continuing and going on after we die. To some this is a comforting thought, to others unnerving and alien.

There is also a thought experiment that we can perform that is suggestive of dualism. Imagine that I wake up, as usual, and open my eyes, or think I do. To my left I see my cup of coffee in a clean white mug, steaming a bit and smelling good. "Great," I say to myself, "time for coffee." I glance down the bed, and I am surprised, because it looks unrumpled and perfectly made, as it was the night before. Things begin to get even odder when I notice that where my feet should be sticking up under the covers, the cover is completely flat. The next odd thing I see is that my torso also does not make a lump under the covers. When my wife pulls back the covers and asks whether she can hand me the coffee, I become most alarmed: I see no body at all where my body should be. Is this a nightmare? No, I am fully awake, but my body seems to have vanished in the night. It is not merely that it is invisible. It simply isn't there. It has disappeared. Perhaps it no longer exists. Have I turned into a pure consciousness? What philosophers call my mind or consciousness (though these most certainly are not the same thing), including my thoughts and visual and tactile sensations, and all the other sensations, of motion and action, is still there, unchanged. I still have the chronic feeling of pain in my back, where my back ought to be, but seem to be missing the back itself. What am I supposed to think? It seems natural to say that my body has gone, but that my mind is still there. I now see that my mind and my body are distinct, then, for my mind can exist without my body.

I can imagine all this; and, more importantly, it is all possible, in the sense that a story of waking up without a body does not seem to be a contradictory story. Freedom from contradiction, rather than imaginability, is the proper test of possibility. If there is no contradiction in the description of an event, then the event is possible. Suppose that it is possible that I shall win the lottery. I can imagine that I shall win it, but that is not the important thing. The important thing is that I can describe myself winning tomorrow, going to the office of the lottery, presenting my lottery ticket, picking up my winnings, and so on, and among the descriptions of all these happy events there is no contradiction. Imaginability *may* be an indication of describability, but it does not guarantee it, whereas describability—in the sense of description genuinely free from contradiction—does demonstrate possibility.

By this test we should conclude that it *is* possible that the mind and body should exist without one another. It is possible that I should wake up with my mind and consciousness intact and my body gone. That possibility is the central claim of dualism. It is significant that the initial claim is not that they *do* exist without one another, since for the moment anyway they are somehow stuck together, but that they *can*. If they can exist independently, it does not follow that both do, or that they will not exist at the same time. We can imagine that mind and body can exist independently of one another, but that at death both of them get destroyed at once, though by two different sets of forces, one physical and one nonphysical, assuming we can make sense of the idea of nonphysical forces.

The main difficulty for interactionism is one that stumped Descartes. *How* can the mind and body interact, if one is physical and therefore spatial, and the other is non-physical and therefore nonspatial? Of course, it is possible to deny that the mind is nonphysical, and I will discuss this important option in the next chapter. But for the moment we are considering the view that the mind is nonphysical and the body is physical, and that the two interact. The question is, *how*? How can mind and body interact if one is physical and the other is not, given that physical and non-physical things cannot interact? This is the objection to interactionist substance dualism made by Princess Elisabeth and Gassendi, and it is hard for interactionist substance dualism to meet it. Perhaps it is impossible.

There have been some contemporary attempts to make dualism work, but on the whole they have been a bit disappointing. E. J. Lowe, for example, argues for what he regards as a new picture of psychophysical dualist interactionism.[1] He notes that the structure of the causal chains of events is a branching structure, but since the chains get intertwined the structure as a whole has no ends. So mental events cannot interact or indeed fail to interact with the tips and initiate causal action, since there *are* no tips! When I make to lift my arm, the tree structure as a whole is activated, or some significant part of, but it is as a result not of this but of the desire or wish or intention to move my arm that my arm moves. The activation of the neurophysiological causal tree explains the exact *way* in which the movement of my arm occurs, say, jerkily or smoothly, but it does not explain *that* it occurs in the first place. The tree "mediates" the relationship between mind and action. But mediation is a causal relationship. It remains true that what explains the fact that I raise my hand is the decision to raise it. So there is a direct action of the mental on the physical that still needs explaining. Lowe claims that my mind communicates not with the tips of the tree of causal events in the brain, since there are none, but with the whole tree, and explains the existence of the whole tree-structure of neurophysiological events. But the problem is just the same. How does the mind activate the whole tree? If interacting with the tips of the tree is impossible, so is interacting with the tree as a whole.

Another odd feature of this account is that the intention and the activation of the tree of events take place at the same time, the one responsible for the *how* of the movement, and the other responsible for the *fact* of it. This seems fishy.

Moreover, Lowe's version of dualist interactionism also does not eliminate the charming "pairing problem" that arises for dualism.[2] My mind issues the wish for my arm to rise, and the wish instructs my tree-structure of events to begin. Suppose you are standing next to me. How is it that my mind doesn't go into the wrong tree structure—yours—and not mine? Or how come it doesn't go into both, like a radio broadcast? If mind and body are genuinely distinct, then how is my mind paired with my brain and your mind with yours? Why does my arm rise and not yours?

Property Dualism

There is an answer to the pairing problem, but it means abandoning substance dualism. For dualists who are daunted by the problems facing substance dualism, another kind of dualism may seem to afford some relief. It is called *property dualism*.

The property dualist sees clearly the difficulties of two interacting but distinct substances, and proposes instead a dualism not of substances or *things* but of their *properties*. There is only one substance, says the property dualist, but it has two sorts of properties, physical and nonphysical. In one version of property dualism—the physicalist version—the mind is physical. It is the relevant part of the brain for causal interaction. But it has two sorts of properties.

Consider the fact that a piece of art such as a painting is physical, but it can be said to have nonphysical properties. Though it is made of paint and wood and canvas, the painting can be said to be: accurate; a bit of a caricature; witty; slightly derivative though stylistically effective; and a bit dark. These are aesthetic properties, not physical ones. But there are not two things or substances, a canvas *and* a work of art. If someone attacks a painting with a knife, as people sometimes do, then it might lose some or all of its aesthetic properties. And we cannot say that the higher-level properties of *being witty* or *being a bit dark* are identical with the paint and wood and canvas, though they are dependent on them.

Mental properties, such as having a thought, are *grounded* in the physical brain or mind, says the physicalist property dualist, but they are not themselves *reducible* to physical properties. If my brain is damaged, my capacity for thought can be impaired. But according to the property dualist, it does not follow that mental properties, such as my having a thought, are physical.

Now the property dualist is in a position to respond to the pairing problem that attacks substance dualism. Why do *my* mental activities, if detached from my body, not cause things to happen in *your* body? How do my mental activities reach the correct destination? Why is *this* mind connected with *this* body and not some other?

The property dualist denies that the mind and the body are distinct, since the mind is a physical thing. Mind and body can then interact nicely. Though physical and nonphysical substances cannot interact, the mind is not nonphysical; it is physical. But it *does* have nonphysical properties. These properties, however, do not themselves have effects on the body.

Everything seems to be in order. But there is a large fly in this ointment. Although it may be true that abstract triangles and aesthetic properties do not have actual effects, in the case of the mind, mental properties, such as thoughts and feelings, most certainly do initiate effects. Intent, premeditation, or *mens rea*, the "guilty mind" which is presumably something mental (since "mens" is just the Latin word for mind, and from which the English "mind" and "mental" are derived), is an element that is necessary to demonstrate certain classes of crimes, most notably murder. The evil intentions in the mind are taken to be the properties of the guilty mind that result in the unlawful deed. And the physicalist property dualist has no way to account for these.

Parallelism

There is another dualist possibility, however: parallelism. On this view, mind and body are distinct, but they do not interact. We can accept dualism, including the proposition that the mind and body can exist independently of one another, along with the proposition that one is physical and the other is nonphysical, and at the same time reject the proposition that they interact. One good reason we can give is that "Physical and nonphysical things *cannot* interact," and we have seen exactly why this is such an appealing proposition, starting with the arguments offered by Princess Elisabeth and Gassendi. We simply *deduce* "Mind and body cannot interact" from "Physical and nonphysical things cannot interact," given the further premises that the mind is a nonphysical thing and that the body is a physical thing. We arrive at:

(1)   The mind is a nonphysical thing.
(2)   The body is a physical thing.
(4)   Physical and nonphysical things cannot interact.

As solutions go, this one is as logically appealing and as successful as any other. From (1), (2), and (4), it certainly does follow that mind and body *cannot* interact, and hence that they *do not* interact. But how then are we to account for the appearance that they *do* interact? It strains belief to suppose that they do not, one might think, because the fact of mind–body interaction is so common and familiar as to be undeniable. There is the effect of alcohol on mental state, for example, not to mention drugs of various sorts. There is the effect of the mind on the body, most obviously of cases of intentional action, but also in cases, say, of a mental state such as blind rage leading to unhappy physical consequences. Psychiatry and psychology are full of examples of interactions, in both directions.

What does parallelism have to say to all this? It is perhaps surprising, but these examples represent no threat whatsoever to the parallelist view. The parallelist can simply assert that though there is no *interaction* between mind and body, there is a *correlation* between what happens in the body and what happens in the mind wherever we thought there was an interaction. The drinking of beer is *followed* by the fogginess of the mind, or correlated with it. And this is a well-established empirical fact that is neutral with respect to interactionism and parallelism. What we must not do, says the parallelist, is to imagine the body emptying beer into the mind, or, what is equally absurd, getting the neurons to fire into the mind, or in some literal sense sending physical messages directly into the mind, so that we have the ridiculous picture of electrical signals going off in the mind as well as in the body. We have no way at all of picturing such an event, as the mind is nonphysical and the signals are physical. We would then be imagining something that does *not* have a position in space (the mind) containing objects that *do* have positions in space (signals from the neurons). As A. J. Ayer has observed,

The physiologist's story is complete in itself. The characters that figure in it are nerve cells, electrical impulses, and so forth. It has no place for an entirely different cast, of sensations, thoughts, feelings, and the other *personae* of the mental play. ... Nor are there such temporal gaps in the procession of nervous impulses as would leave room for mental characters to intervene. In short, the two stories [mental and physical] will not mix. It is like trying to play *Hamlet*, not without the Prince of Denmark, but with Pericles, Prince of Tyre. Each is an interpretation of certain phenomena and they are connected by the fact that, in certain conditions, when one of them is true, the other is true also.[3]

The impossibility of physical and nonphysical things *interacting*, asserted in proposition (4) of the initial tetrad, does not prevent the *correlation* of the events within the physical body and the nonphysical mind. What the parallelist objects to is the idea that the electrical impulses or neural activity do in any literal sense sidle right up alongside the mind, and, from their close proximity, interact. There can be no literal proximity to the mind, if "literal proximity" is spatial contiguity.

In the history of the mind–body problem, parallelism arose partly as a result of a vivid awareness of the reasons for which Descartes's interactionist dualism could not work. The *dualism* made the *interaction* impossible, and behold: the mind–body problem was born. The parallelists were committed to dualism. So what was left? Mind and body *could* not interact, just *because* dualism was true, they thought. But mind and body *do* seem to operate in tandem—synchronized, as it were. When the desire for coffee enters the mind, it is then that the body, or part of it, reaches out and picks up the cup of coffee. Then the mind says to itself, "Enough. No more," and the body stops pouring the coffee into its throat, and puts the cup down. But why does it do it *then*, at exactly that moment? How has the mind *made* it come to pass that the body stops pouring coffee down its throat?

Causal interaction, said the parallelists, just *is* synchronization. The most celebrated and extraordinary metaphor for this idea to appear in the post-Cartesian wave of parallelism in seventeenth- and eighteenth-century France was the image of two clocks beating and striking in synchrony, satisfying the very French desire for order and harmony that existed at that time. The parallelists invite us to imagine two synchronized clocks, keeping perfect time together. When one strikes three o'clock, say, so does the other. If we were to imagine a slight time lag between them, it might seem tempting to think that the clock that strikes first *makes* the second clock strike. That would be a false inference, a fallacy that actually has a name: *post hoc ergo propter hoc*, or *after this therefore because of this*.

Leibniz was the most celebrated of the parallelists in the seventeenth and eighteenth century. He may have been struck by a well-publicized phenomenon observed by his teacher Huygens (himself a pupil of Descartes), who was the inventor of the pendulum clock. Huygens had been ill in bed, and while lying there had noticed that the pendulums of two clocks mounted in one case always ended up synchronized, though in opposite directions ("antisynchronized") irrespective of their starting points, displaying what he called an "odd kind of sympathy." The clocks were somehow regulating one another, but just how remained a mystery until 2002. In that year a team of scientists from Georgia Institute of Technology were able to explain the phenomenon with a sophisticated mathematical and physical model based on small vibrations in the case that interfere with one another.[4] After ruling out air motion experimentally, Huygens had himself suspected but not been able to prove that the phenomenon was caused by small motions in the clock case, and the Georgia team proved him right.

Leibniz went considerably further with the thought, however. Mind and body do indeed act *as though* they were synchronized, and although they do not affect one another in a literal way, for him synchronization *is* causation. Nothing could go into or out of a "substance," in the terminology of early modern philosophy, or a genuinely unified individual thing, which in this respect is rather like an indivisible atom. The individual thing is "window-less," in Leibniz's metaphor. According to him the "mental pendulum" and the "physical pendulum" are synchronized in their behavior—fortunately not antisynchronized—though not by interaction. The synchronization comes with their initial creation by God from the substances' "complete individual concepts," which detail everything that will happen to them throughout their futures. This view, of course, has implications for freedom of the will, in which Leibniz was keenly interested. What concerns us, however, is the fact that the whole universe is arranged so that what we observe of it manifests all sorts of remarkable synchronizations. These include but are not limited to the synchronizations of the mind and the body, which, along with all the other synchronizations that constitute the laws of universe, are designed to bring about the best possible universe over time.

There is a big difference between Leibniz's views and those of the so-called occasionalists. The occasionalists took the view that parallelism is true, but that physical events in the body are the *occasion* for God to act in the mind, and vice versa. The occasionalists, such as Geulincx, who before Leibniz had used the simile of the two clocks to illustrate parallelism, were impressed by the absolute power of God, and wanted to make all our actions and every other action in the world into actions whose motive or moving power is God. That this uncommonsensical and implausible view survived as long as it did is a testament to the religious faith of the time, and to the dedication of the occasionalists in following their reason through to where it led, or seemed to lead. On the other hand, as Leibniz pointed out, the continuous need for action on God's part every time mind and body interact makes for a very hard-working God, and it is itself unacceptable on religious grounds as well as the ground of philosophical economy and the theological drive toward simplicity and piety.

The Role of the Conservation Laws

Another historically important point about interaction needs to be made on behalf of those who, like the parallelists, wish to deny that mind and body interact. It has to do with the laws of conservation in physics. Among these laws, which seem to be about as well established as anything could be in physics, the conservation of mass and energy tells us that in a "closed" system changing over time, the net total of mass or energy in the system stays the same. The system as a whole neither gains nor loses mass or energy. (There are particles with no mass, but they must have some energy, since energy is a function of frequency.)

Suppose that the human body is a closed physical system. In other words, it acts as it does because of the physical energy and mass that it contains, and it is insulated from the effects of outside energy. This has been called "the causal closure of the physical." If we want to change anything within the system, we will either have to use the energy that is already within the system, or we will have to introduce energy from the outside. If we use the energy in the system, then the mind, since it is not within the body, can have no effect on the body. If we do not use the energy already in the system, then mass and energy are not conserved, or the system is not closed.

However, if the mind is to effect a change in the body, then it must presumably introduce physical energy into the body. But according to our first proposition, the mind is nonphysical, and so it cannot expend physical energy. Here we can see that the conserved mass and energy are playing the same role as linear dimensions did in our first formulation of the mind–body problem. Lack of linear dimensions and spatial location on the part of the nonphysical is what makes the physical and nonphysical unable to interact. But the same result is obtained if we make mass or energy the defining characteristics of physical things. Physical and nonphysical things cannot interact. The body will not accept nonphysical energy, and the mind will not accept physical energy, in both cases because of the causal closure of the physical.

Versions of the four propositions are often if not always there when the mind–body problem is discussed. A specialized form is present when conservation laws are at issue. Ernie Lepore and Barry Loewer,[5] for example, express the mind–body problem as the difficulty of fitting together the following three propositions:

(5)   The mind and the body are distinct.
(3b) The mental and the physical causally interact.
(4b) The physical is causally closed.

Roughly speaking, (1) ("The mind is a nonphysical thing") and (2) ("The body is a physical thing") give us (5), proposition (3b) works like proposition (3) ("Mind and body interact"), and (4b) implies (4), that the mental and the physical cannot interact. The problem is that the physical world cannot reach out of itself into anything else that is nonphysical, but it must somehow interact with the mental, which is nonphysical. Similarly, the mental world cannot reach out of itself into anything else that is nonmental, but it must somehow interact with the physical, which is nonmental. Describing the inability of the mind to reach into the physical and of the body to reach into the mental is a way of stating the existence of the law of conservation of mass and energy, which has (4b) as a consequence. "Causally closed" means that energy or mass from causes outside the physical world, or outside the closed physical system, cannot get into it, and that it cannot contribute energy and mass to outside and nonphysical systems in such a way as to deplete the net total of its own energy and mass.

Naturally, if the mind is physical, then the body plus the mind can function as a closed system, and there is no difficulty with the laws of conservation. This amounts to denying (5), that the mind and body are physically distinct, which they are not, according to the physicalist.

The only option that does not seem available, given what physics has to say about conservation, is the denial of (4b). So a physicalist will deny (5) and affirm (3b). This involves the interesting claim that the mental is physical, or a denial of the claim that the mind is nonphysical. A parallelist, on the other hand, will affirm (4b) and deny (3b), telling us that the mind and the body are indeed distinct, but that they occupy parallel and noninteracting realms.

The lesson so far is that we should be either parallelists or physicalists, but not interactionists.

Epiphenomenalism, Emergentism, and Supervenience

There is another form of dualism that was especially popular at the end of the nineteenth century. It has seen a modest resurgence recently, in much more sophisticated forms, though more as an object of study, perhaps, than as a view actually to be believed. Known by its formidable Greek-derived name, *epiphenomenalism*, it is the claim that mental events and the mind are "epiphenomena." "Epi-" is a Greek prefix that means "on the occasion of" or "in addition" to. "Phenomena" are the things that appear, or happen, so epiphenomena are things that appear in *addition* to what might be called the basic phenomena. For most epiphenomenalists, if not all, the basic phenomena are those of the physical world, and mental phenomena and events are attendant on physical phenomena. Epiphenomenalism is the view that physical events cause mental events but mental events do not cause physical events.

There is an obvious comparison to be made with shadows. My hands curled up in the right way can be made to cast a shadow that looks like an eagle's head onto a wall or screen. The shadow is dependent on my hands, but what my hands do is not dependent on what the shadow does. It would be amusing but physically difficult for the eagle on the screen to open its beak, say, and force my fingers to move. The image of the eagle projected onto the screen is *just* a shadow.

Almost nobody holds or has held the reversed epiphenomenalist view that mental events cause physical ones but not the other way round, and it is, I think, fairly obvious why. For one thing, there are the obvious phenomena to think about, such as brain damage. But at a deeper level the epiphenomenalists are those physicalists who want to be strict physicalists but who cannot quite see their way to deny the existence of fully mental events, though they also find it hard to see how mental events can exist at all. In

Epiphenomenalism is the view that physical events cause mental events but mental events do not cause physical events.

conformity with their physicalism, they then downgrade the importance and causal power of mental events as far as possible in the physical scheme of things.

Still, if epiphenomenalists are really physicalists under the skin, they are inconsistent ones, since epiphenomenalism admits the existence of genuinely mental events. There are mental events, it claims, but they have no causal power, unlike physical events.

According to the Victorian biologist Thomas Henry Huxley, our consciousness is a "collateral product" of the "mechanism of the body" and "as completely without any power of modifying the working of the body as the steam-whistle which accompanies the work of a locomotive engine."[6] Volition too is an emotion that "indicates" physical changes but does not "cause" them.

There is something clearly wrong with Huxley's simile of the steam-whistle, since nothing prevents us from rigging up a steam-whistle so that every time it blows, the steam activates a fan rigged to an electrical circuit that brakes the train, and that when the whistle is not blowing, the train resumes its normal speed. The whistle then has definite and specifiable physical effects, and there is nothing in the nature of the physical world to prevent this sort of causal loop.

In the case of the mind or consciousness or soul, Huxley would rule out the causal loop. Yet why is it impossible? *Why* is the mind causally inert? Huxley does not address the question, but it certainly seems to push epiphenomenalism hard in the direction of property dualism. For it is an odd thing indeed, an odd substance, that can have no effects whatsoever. Properties seem more suited than substances to causal inactivity. Even so, one might think that the property of *being hot* can cause me to have the property of *wanting a drink*. Why do I have the property of *wanting a drink*? Because I have the property of *being hot*.

We should keep clearly in mind the fact that epiphenomenalism is a form of dualism. It allows interaction between mind and body in one direction, from mental to physical, but not the other. But there is still a contradiction here. Epiphenomenalism has cut the mind–body problem down to half its original size, so to speak, but what remains is every bit as intractable as the original full-scale version. We don't have to deal with the mind acting on the body, but how can the body act on the mind, if the mind is nonphysical, and physical things cannot act on nonphysical things? In fact I think epiphenomenalism counts as a rough-and-ready philosophy of mind, but not as a genuine solution to the mind–body problem. This may explain why philosophers have on the whole been less than interested in it, and why it has been referred to as "the curse of epiphenomenalism" by one writer (Stephen Law) in the philosophy of mind.[7]

Emergentism is a view of the relation between mind and body of roughly the same kind as epiphenomenalism, in the sense that the physical is dominant and the mental is a sort of by-product, but it is important to see what the difference is between the two. Epiphenomenalism is a kind of dualism, in which two separate kinds of events exist and are causally related. With emergentism, the relation between the mental and the physical is much closer. It should perhaps be discussed later, in the next chapter, as a form of physicalism, but it seems to me that the comparisons and contrasts between epiphenomenalism and emergentism are interesting ones.

There is a mystery, very much at the center of the modern mind–body problem, of why it should be *pain* that emerges from the brain areas that are activated by $A\delta$ or C fiber stimulation. (The $A\delta$ fibers are associated with acute and sharp pain, the C fibers with dull or burning pain.) But, according to the emergentist, there simply is *no answer* to the question why it is *pain* that ultimately emerges from the brain areas that are activated by $A\delta$ or C fiber stimulation, and not something else entirely. Pain does not emerge from the stimulation of the fibers in the way that sixteen ounces just turns into one pound; but then one wonders how on earth it *is* related to the stimulation of the fibers. Well, it *emerges* from them, comes the answer.

The emergentists both accept and deny (1), that the mind is nonphysical. On the one hand, the mind is physical, because it is really driven by the structures from which it emerges. On the other hand, the mind is nonphysical, because it has "emerged" from the physical. But how can truly novel properties, such as pain, emerge from the stimulation of the fibers? If they "emerge" *from* the physical, then they are nonphysical. But if they are genuinely nonphysical, how can they "emerge," and why do they need to?

It seems to me that the emergentists must make up their minds. If with the mind we have a genuinely new phenomenon, a nonphysical and nonspatial one that has emerged like a butterfly out of the chrysalis of matter, then it cannot affect the body, since the body has exclusively the wrong kind of properties to interact with the mind, that is, physical and spatial ones. From this point of view it is hard to see how mind could emerge in the first place, since in emerging it makes itself spatial. If, on the other hand, the new phenomenon has a complete dependence on the physical and spatial phenomena, and can engage with them, it is hard to see how it is anything *but* them, and therefore not a new and emergent property at all.

Emergentists accept the fact that mind can turn around and *do* things to matter, but they do not explain *how* this can happen if the mind has "emerged" and is not physical. If the mind has emerged as nonphysical, we need to understand the way in which it can then interact with the physical. And that is the mind–body problem.

There is a concept that may seem to help with understanding how something can both be nothing but its base properties, and at the same time something emergent, something in addition to the base properties. Many, even most, emergentists have used the concept of *supervenience* and been grateful for the light it casts on the relation between mind and body. According to these emergentists the mind *supervenes* on the body.

The concept of *supervenience* is a difficult one, but the main idea is something like this. Suppose a property A supervenes on a property B. For example, some geometrical or aesthetic property A supervenes on the properties, collectively B, for "base," of a spatial figure or of painting, such as being thus-and-such a closed figure, or having thus-and-such colors, lines, and forms. We say that A supervenes on B when there cannot be a change in A without there also being a change in B. One cannot suppose the aesthetic properties of the painting changing without the physical properties having changed as well. If the aesthetic properties are to differ, so too must the physical properties. In this sense it can be said that the A-properties are generated by the base properties.

The emergentists who make use of the concept of supervenience believe that (2) the proposition "The body is physical," but will reject (3), the proposition "Mind and body interact." Instead they will say that mind *supervenes* on body, or more particularly on the part of the body with the right kind of tissue, namely, the brain.

And yet there does seem to be a kind of causal power possessed by the human mind and consciousness, recognized in (3). Mind interacts with body, which is to say it has effects in the physical world. Emergentism, even with the more refined concept of supervenience on board, cannot do justice to this causal power. The reason is that emergentism is actually a form of physicalism, and it attempts to deny the existence of the nonphysical except in a very diluted form that cannot accomplish what philosophers call "mental causation."

The reason emergentism is not a very popular view is that it is not a very clear one. On the one hand, the mind "emerges" and engages in mental causation. On the other hand, it is the creature of the forces from below on which it supervenes, and cannot attain to any sort of causal power.

PHYSICALIST THEORIES OF MIND

Behaviorism

Given the troubles of dualism, one may be tempted by what is easily the most straightforward solution to the mind–body problem: physicalism. On this view, everything that exists is physical; so the mind is a physical thing, if it is a thing. If proposition (1), that the mind is a nonphysical thing, is false, which it is if everything is physical, then the mind–body problem is solved. The mind is a physical thing, and so there is nothing to stop it from interacting with other physical things, including the body. It remains true, however, that physical and nonphysical things, on this view, cannot interact. But it doesn't matter, since there are no nonphysical things.

Well and good, but *in what way* is the mind supposed to be a physical thing? There are a number of different possibilities.

Behaviorism is the view that the mental is the behavioral. Mind is behavior. The mind is the body, considered from the point of view of its behavior. Some hardline behaviorists actually went so far as to deny the existence of the mind and mental events, over and above behavior. There is no mind, but only behavior. This is a very simple but pretty extreme point of view that has not found much favor among philosophers or scientists recently. Part of the problem is that we do seem to be acquainted with our own mental states, our thoughts and feelings, and they are not nothing at all. Another part of the problem is that there do seem to be obvious examples of an interaction from mind to body.

A second and more reasonable version of behaviorism took the line that, from a scientific point of view, we should not study the mind and mental events, because they cannot be directly observed; their existence must be inferred from the external behavior of human subjects. This is not the strongest line of thought, it has to be said, since many entities studied in science cannot be observed directly, but we infer their existence from their effects. Electricity is an example. We know about it by watching lightning, for example, or by understanding Maxwell's equations, or how a radio works.

Nevertheless, one can understand how, in the atmosphere of the religiously oriented dualism that prevailed in philosophy at the beginning of the twentieth century, and which many scientifically oriented people found un-congenial, the bold claim could be advanced, on behalf of psychology, that science should allow as its subject matter only what can be directly observed. This is certainly very different from saying that its subject matter does not exist.

An even more reasonable variant of behaviorism is that mind as such is not interesting or important, and its study should be replaced by the study of behavior. There is no mention in this view of what is directly observable. It is almost like saying, "I am more interested in behavior than I am in mind." This is, of course, an impossible view to rebut, if it is true that you are more interested in behavior than in mind; but the question remains whether you *should* be more interested in mind as such, or whether its study would offer you some benefit.

This third and more reasonable line of thought, how-ever, would make it impossible to solve the mind–body problem in a way that is satisfactory for science, or even to *state* it. We should not study or talk about minds, so we will never be in a position to say either that the mind is a nonphysical thing with any scientific authority, or, for that matter, that it is a physical thing. Our first proposition, that the mind is nonphysical, is one whose truth or falsity we should not actively pursue, because its truth or falsity is something that should not be talked about!

There is a fourth form of behaviorism, however, that is more appealing than any of the first three forms. It is the view that the mind is behavior in the sense that any proposition about the mind can be "translated" into a proposition about behavior. So, for example, if I say "I am tired," I am reporting not the presence of an inner feeling of drowsiness, but rather of a tendency or disposition to stop work, to lie down, to close my eyes perhaps, to rest, and so on. All of these things are external behavior, observable by others and fully within the purview of science and of common observation.

Gilbert Ryle wrote in his influential 1949 book *The Concept of Mind* that

> when we describe people as exercising qualities of mind, we are not referring to occult episodes of which their overt acts and utterances are effects; we are referring to those overt acts and utterances themselves.[1]

It is hard to believe, reading the admittedly rather few passages like this in his book, that Ryle was not a behaviorist, and indeed he himself remarked of the book that when he wrote it, "certainly one of my feet was pretty firmly encased in this boot."[2] Nevertheless, there is more to the story. Ryle writes in the passage above that when we talk about minds, "we are not *referring* to occult episodes" (my emphasis); but there is a case to be made that all the same he does not deny the *existence* of these episodes. Perhaps he means that when we say publicly that a person is tired, we are "referring," not to that person's private and inner feeling of tiredness hidden from others, but rather to his tendency or disposition to stop work, to lie down, to close his eyes, to rest, and so on. This is not to deny that the inner feeling exists. In chapter 6 I describe the other side of Ryle's view, his "dissolutionism" as it has been called, and again take up the question whether he is to be considered a full-blooded behaviorist.

What is wrong with the idea that the mind just *is* some sort of behavior? One difficulty is that this view seems to leave out what we think of as the "inner" life of thoughts and feelings—the mind! Behaviorism solves the mind–body problem by denying the mind in one way or another. We can produce behavior without it, and without its rich experience of sensation and perception, colors, sounds, and tastes, for example, or qualia. We can easily imagine a machine that reacts to red things just as we do, picking them and eating them, perhaps, but which has no experience of the colors. It behaves as if it saw red, but it does not have the experience. This has been called the "problem of absent qualia."

There are other twentieth-century philosophers who, like Gilbert Ryle, might also seem to be offering behaviorist arguments but are not. A famous example is the "beetle-in-the-box" part of the so-called private-language argument in Wittgenstein's *Philosophical Investigations*. Wittgenstein is arguing that there could be no language in which we could report our own private sensations. Suppose, he writes, that everyone has a box with something in it, or perhaps nothing at all. There is a rule that no one is allowed to look inside anyone else's box. Everyone calls what is in his own box a "beetle." But where no checking is allowed about what is in anyone else's box, the word "beetle" would not come to mean "an organism of the order *Coleoptera*, with hard fore-wings," but rather "whatever is in anyone's box." Yet Wittgenstein explicitly denies that he is trying to deny the existence of sensations, somethings in the boxes. The issue is one of meaning.

Behaviorism does indeed solve the mind–body problem, very easily, by denying that the mind is a nonphysical thing. Behaviorism simply denies proposition (1). So the discussion at this point should turn to the question of how plausible behaviorism itself is. The judgment of history, it is fair to say, is "Not very." One powerful reason is the problem of absent qualia, mentioned above. Another objection is the possibility of the *inverted spectrum* and its analogues in other sensory modalities. We can imagine people behaving systematically in the right way, but having the "wrong" experiences. Their "inner experience" might be of all the colors, but with their positions in the visual field reversed

Behaviorism does indeed solve the mind–body problem, very easily, by denying that the mind is a nonphysical thing.

from ours. The subjects with inverted color experiences would see a cyan green-blue color where we see red, a blue where we see orange, and so on throughout color space. But the *behavior* of these people would be the same as ours. When we see red, and call it "red," they see what we call "cyan," and call it "red," and when we see cyan, and call it "cyan," they see red and call it "cyan." Accordingly, having the experience of red cannot be a matter of producing the right behavior. Our subjects suffering from an inverted spectrum *behave* around red just as we do, even calling it "red," but actually experience a green-blue cyan color. According to behaviorism, the subjects are *experiencing* red; but this is false. Therefore, behaviorism is false.

There are other overwhelming arguments against behaviorism, but perhaps the biggest has been the realization from psychiatry, psychology, and physiology that events in the brain can explain behavior. If the relevant parts of the visual cortex are absent or damaged, for example, color vision can be affected, and our behavior will not be the same as the behavior of someone with a properly functioning visual cortex. During the two World Wars the evidence from neurology and from the hospitals mounted up. It began to look as though the state of the brain is what is making us behave in the way we do, or at the least allowing us to—though these are hardly the same thing. When in the 1950s the evidence for a causal explanation of behavior in the brain, or anyway a causal explanation of abnormal behavior, became too telling, behaviorism started to lose almost all its popularity, and very quickly at that.

It was especially troubling that if, according to behaviorism, a mental state is a disposition to behave, then if what explains the behavior is the mental state, as we would ordinarily think, *we have to say that what explains the behavior is the disposition to behave in that way!* Thus behaviorism amounts to a tautology—a trivial truth—if there is such a thing as an explanation of the body's behavior by mental causes.

The Identity Theory

By the mid-1950s, when things began to change, they changed completely. Starting with a pioneering paper in 1956 by U. T. Place, more and more philosophers and scientists were persuaded that the explanation both of what people do and of what they experience lies in the brain. American and Australian philosophers in particular began to advance what became known as the "mind–brain identity theory," or the "identity theory," as it is called for short. This view, as its name suggests, is the claim that mind and brain, or anyway the relevant bits of the central nervous system, are identical, one and the same. Here too, the mind–body problem is solved at a stroke, by physicalism, by the denial that the mind is a nonphysical thing. Every mental event is a physiological event within the nervous system. Accordingly, the theory that the mind is the brain has sometimes been known as "central-state" materialism, a materialism making the mind into the central nervous system, distinguishing it from the "peripheral-state" materialism of the behaviorists.

In its favor, the theory can be said to be commonsensical, given the facts of neurology such as the effects of brain damage, and it makes a great simplification in the philosophy of mind. But it is hardly an "astonishing hypothesis," as Francis Crick claimed in a book of that title published in 1994. It is important and interesting, certainly, but not so astonishing. Like behaviorism, it solves the mind–body problem at a stroke, by denying that the mind is nonphysical. If this proposition about the mind is true, then the solution is, as before, impeccable. The mind is the brain and the brain is a physical thing, so the mind can interact with the rest of the body without difficulty. Yet we miss the essential thing needed for a solution: how has the physical, which has physical properties, turned into the mental, which has properties incompatible with being a part of the physical? What do neurons have when they fire that produces mind rather than electrical signals, or soap bubbles, for that matter?

Against the theory are also certain logical and philosophical difficulties. The central-state materialists do not claim and are bound not to claim that the word "mind" *means* "brain," which is fortunate for them, as "mind" as a matter of fact does not mean "brain." If it did, the claim about the meanings of the words would make the main claim of central-state materialism (that the mind is the brain) into a necessary truth generated by the meanings of the two words. Its truth could have been discovered simply by looking in the dictionary. However, what the mind *is* was taken by the central-state materialists to be an empirical and factual question, not one of meaning. Central-state materialists, including Crick, took the question to be scientific, in just the same way as the question of what the gene or unity of heredity is was empirical and factual, to use the central-state materialists' own favorite example. The gene turned out to be DNA, but this could not have been known from the meanings of words "gene" and "deoxyribonucleic acid."

So far so good. But then there appeared an unpleasant proof from the world of logic. Identity, as it turns out, is always necessary. Suppose $a = b$. a has the following interesting property. It is necessarily identical with itself, a. Take this last statement, that a is necessarily identical with a. Substitute b for the second a; we are entitled to do this, since we have supposed that $a = b$. But now it follows that a is necessarily identical with b. Accordingly, if central-state materialism is going to claim that the mind and the brain are not necessarily identical, it must itself be false. This proof was published by Saul Kripke in lectures given in

1970, and he developed extraordinarily interesting related arguments in the same work.[3]

Proofs of this sort, it should be noted, rely on the fact that the terms on either side of the identity sign, here "a" and "b," are fixed names ("rigid designators," as Kripke called them) and not descriptions that can be applied to different things. "The human mind" and "the human brain" are names, and so are "pain" and "events a in the thalamus, b the pre-frontal cortex, or c the primary and secondary somatosensory cortex (S1 and S2)."[4] So the proof does not imply that it is somehow necessary that the Queen is Elizabeth II, which is true as I write. "Elizabeth II" is a name, but "the Queen," even "the Queen of England" is really a compressed *description* that can apply to different persons, as it has done in the last hundred years. It is not a rigid designator because the place of the object of its description can be different objects.

Furthermore, the claim that the mind is the brain also turns out to be equivalent to the claim that the brain is the mind, since identity is what logicians and mathematicians call "commutative." If $a = b$ then obviously $b = a$. But the claim that the brain is really at bottom the mind could hardly be expected to appeal to a hard-headed central-state *materialist*, since it makes a claim more suggestive of *idealism* (everything is mind) than of materialism (everything is matter).

What is a central-state materialist to do?

One answer was to take advantage of a distinction that had existed for some time in general philosophy, including metaphysics and the philosophy of art: the distinction between *types* and *tokens*. Take, for example, Edward Elgar's Cello Concerto in E minor. It has been played many times, including its disastrous premiere in 1919, Jacqueline du Pré's triumphant and elegiac performances in the 1960s, and hundreds of others. How many Elgar Cello Concertos are there? Could one say that there are hundreds? In that case, since Elgar wrote the work or works, he wrote hundreds of Cello Concertos. But he didn't. He was enormously hardworking, but not that hardworking. Or is there only one concerto? But then how could it appear in all sorts of different places and at all sorts of different times with so many different soloists? The answer developed by philosophers is that there is one concerto *type* and many concerto *tokens* or *instances*, in much the same way that there is one book called *Pride and Prejudice*, but many *copies* of the book. The copy both is and is not the work; it is a token of the work, but it is not the type. There is a difference between the Cello Concerto case and the case of the book, though, because there is nothing that could be regarded as the performance of *Pride and Prejudice*. But though what is played is "the music," as it is written, all the same it can be said that the glorious sound that is the Cello Concerto is not the sheet music, whereas the printed copies of the book are the novel.

The distinction has its difficulties, clearly, but it was used advantageously to distinguish two forms of the identity theory. There is the type *pain*, and there is the individual *pain* that is a token of the type. In the stronger and less plausible form of the theory it was the type or property *mental state* that was said to be identical with the type or property *brain state*. In the less sweeping and more convincing version, it was instead said to be just the one particular instance of a mental state that was identical with a particular instance of a brain state. It might be that two organisms both feel the same or a similar pain, but that they are not in the same brain state. They are in some brain state; and since it is implausible that everyone's physiological and psychological systems work in the same way, especially when we consider different organisms that have very different kinds of brains, it is much more plausible to identify *this* pain with *this* brain state, and accept the consequence that two individuals in the same psychological state may not be in the same physiological state. But they must be in *some* physiological state, with which the pain state is identical. So one is bound to wonder what makes all the tokens into tokens of the same type. Why are they all instances of pain?

In any case, it was suggested that the logical arguments against central-state materialism only worked against identities of types. That turned out not to be the case. The arguments, as it was soon realized, worked equally well against identities of tokens.

Even before the logical proofs against central-state materialism were worked out and made public in the 1970s, however, it was already too late; central-state materialism was dead in the water. This came about not because of the intricate logical argumentation against it, but because a much more powerful view had arisen to take the place of central-state materialism, more in keeping with the science of the time.

Functionalism

The new view that took the place of central-state materialism was *functionalism*. It came upon the philosophical scene in 1967 with Hilary Putnam's "Psychological Predicates" and other subsequent papers.[5] Putnam argues that pain is not a brain state, but another kind of state entirely. It is a state of a probabilistic automaton or a *Turing machine*. A Turing machine is in essence a computer, and it computes, having *computational* or *functional states* that are not its physical states. They are described completely differently, for one thing, and for another the computational states are not made of matter, but rather of a kind of functionality, if they can be said to be made of anything at all. One can also imagine that two Turing machines could happen accidentally to be in just the same physical states, but in the process of performing different computations. So their computational states at that moment at which they are physically identical would not be the same states. So if mental states are computational states, as functionalism suggested, they are not the physical states of the organism.

The power of functionalism came from the interesting fact that it deployed to full effect the distinction between computer hardware and computer software. What is going on with functionalism is that the mind is compared to active software, not to rigid hardware. Even with ordinary computers, one can imagine that two laptops computing the same function, say, the multiplication 7×9, might do it in very different physical ways. One might even consider an optical computer that does not work in the same way as an electronic computer, by electrons slowly pushing one another around through the different gates that make up the central processing unit. Clearly the two computers, optical and electronic, are not in the same physical state, since photons are not electrons. But the output (63) will always be the same given the same input (7×9). One can think of the function of the two machines as the same; for even their logical architecture might be quite different. Again, even two electronic computers might be running very different programs yet happen coincidentally at some instant to be in the same physical state.

Putnam had discovered the *multiple realizability thesis*, the proposition that one mental state can be realized in multiple and very different ways. Goats, birds, reptiles, and mollusks all feel pain, depending of course what your philosophy of animal minds is. But it is completely implausible to think that when they do, they are all in the precisely the same physiological brain state.

One might have thought, as Putnam pointed out, that the effect of the development of computers on the philosophy of mind was going to be materialistic, but in the event it was the reverse. The distinction between hardware and software allowed computing systems to be considered in abstraction from their physical states, and to highlight the difference between the computational or Turing-machine state, and the physical.

The time was right for functionalism, and it swept through the philosophy of mind in spite of some rear-guard action by central-state materialists. It rapidly became the preferred philosophy of mind of the artificial intelligentsia, those working in artificial intelligence, but also of many philosophers, especially philosophers of mind, and scientists in fields other than cognitive science.

How does functionalism solve the mind–body problem? The most obvious interpretation is that functionalism denies that the mind is a nonphysical thing, not because it takes the line that the mind is a physical thing, but because it takes the line that it is as wrong to think of the mind just as a *thing*, as it would be to think of a computer program as just another physical thing. What is important about the program is not the physical states of the piece of tape or paper or electronic hardware as a thing, but its functionality.

Putnam has now recanted, arguing in 1991, against his own former functionalist self, that functionalism is false. One of his arguments is that *any* computational description of nonphysical properties can be applied to *any* physical thing, so that functionalism is completely trivial. This is interesting for us, however, whether it is right or not, because it suggests that, before he came to reject it, Putnam had taken functionalism to hold that physical systems, including mental ones, *do* have unique computational descriptions, and that this fact is what is behind the truth that the computational is not the physical. In that case, Putnam must have thought that functionalism solves the mind–body problem by denying the proposition that the mind is a nonphysical thing.

If this is right, it will come as no surprise that functionalism has a problem with qualia or phenomenal properties, just as behaviorism had had. One can easily conceive of two "systems" in just the same functional or computational or Turing-machine state, built into a robot with inputs having spectra inverted relative to one another, and computing on the basis of these inputs. Accordingly, the qualitative experience of the spectrum cannot be the same thing as a functional or computational state. A Turing machine may be computing away without the slightest idea of what it is computing about, the color red, say. It computes, happily accepting inputs and giving outputs about colors, without having the slightest impression or idea what colors are.

An even more interesting idea is that functionalism is a form of property dualism, if it is taken to claim that it is false that the body is physical; for the body, including the brain, might be thought to have nonphysical or functional computational states.

Anomalous Monism

At about the same time as functionalism was changing the world of the philosophy of mind, the philosopher Donald Davidson was independently developing a deep and interesting view of the relationship between the mind and the physical world.

In a classic paper from 1970, "Mental Events," Davidson takes it as given that there are descriptions of events in the world, descriptions that are irreducibly mental, in the sense that they use mentalistic words that cannot be defined by physical terms, as well as physical events.[6] So he subscribes to the essence of the propositions that the mind is nonphysical and that the body is physical, our (1) and

(2). He also takes the view that there are causal relations between at least some mental events and some physical events. But, he notes, such causal relations require a basis in a law that covers them. Where there is a causal relation, there is a law to cover it. There is no such thing as "singular" causation that works on one occasion but not on others. All this is easy enough to accept, until we reflect with Davidson that it is also the case that there are no strict laws covering the relations between the physical and mental events. There is no physical law that absolutely demands that when I decide to go to Italy to see my grandmother, I find my neurons firing in exactly this or that way.

Of course, one might doubt the general truth of this "anomalism of the mental," because there are some pretty strict laws in psychophysics. An example is the so-called Weber–Fechner law for the perception of weight. The law states that in human perception there is a logarithmic relationship between the strength of the stimulus and the strength of response. A correspondingly greater increase in the stimulus is required to increase the same response.

In the first place, however, the law is actually not strict. It applies *moderately* well to human perception, but it only applies well over certain ranges of perception, such as the higher amplitudes in audition or hearing, and there are other limitations as well.

It might well be thought, of course, that though there are no strict psychophysical laws, this is hardly surprising, because there are no strict physical laws either. Gravitational laws, for example, assume a hard vacuum, which never strictly exists. It should be conceded, though, that the laws of physics and chemistry are very much stricter than the laws of psychophysics.

We should also note that Davidson's real interest was in paradigmatically *psychological* laws as they apply to human behavior, or the more rational and conscious parts of human behavior, such as my decision to go to visit my grandmother in Italy, not in perception. There really is no law about such an event or about an event "so described," as Davidson puts it. The intention concerns the rational end of human behavior, and rationality could hardly be codified in such a way as to connect up with the world of scientific law. But our concepts of the mental are tied up with rationality, for example in such ideas as "reasonable," "intent," "intention," "thoughtfulness," and so on.

For Davidson there are causal relations between the mental and the physical, and causal relations demand strict laws, but there are no strict laws between the physical and the mental. And here we have a huge and fascinating problem. We have an inconsistent *triad*, one indeed that has a definite relationship to our original inconsistent tetrad. If the mental and physical interact, and causal relations demand strict laws, then there certainly ought to be strict laws governing physical and mental events. But there aren't, according to Davidson, and we have a contradiction on our hands.

What is actually happening is that Davidson is initially affirming that physical and nonphysical things cannot interact, because that would require strict causal laws between the mental and the physical. He then notes that mind and body *do* interact, but, as he finally puts it, they can only do so under a nonmental *vocabulary*, one that is not physical in nature. What is left is that mental descriptions are anomalous, in that they do not connect systematically with scientific explanations. It is perhaps worth noting that Davidson began his academic career with a PhD in classical Greek philosophy, and that he has always been alive to the richness and variety of language about the mind.

There are things other than mental events that have anomalous descriptions. One might take an interest in things that are *cheap*, for example, without thinking that cheap things have anything in common that could relate them to the physical world by means of strict laws. There are no strict laws of cheapness, if you like. "Cheap" is a vague, idiosyncratic, and interest-relative predicate. It reflects our everyday behavior and practices in such a way that it could never become a word used in a strict science, even economics. And so it is with mental words. They reflect our rational interests, for example in explaining everyday actions, but not within the framework of physical law.

By the way, Davidson also has an argument here for the conclusion that mental events must be token-identical with certain physical events. Since mental events, so described, do not fall under strict laws, and since they do interact with physical events, they must fall under physical laws. Hence they must be physical events. But they must also not be physical events *described* as such, and so they are not type-identical with physical events. So they are token-identical with physical events. This is Davidson's argument that every mental event is actually some physical event. It is certainly a brilliant line of thought.

In his overall argument concerning the mind–body relation, Davidson can be taken to be arguing that:

(1)　The mind is a nonphysical thing

(in the sense that *descriptions* of the mind are couched in nonphysical terms, but not in the sense that it is not a physical object).

(2)　The body is physical.
(3)　The mind and the body interact.
(4)　Physical and nonphysical things cannot interact

(since (4) would require strict causal laws between them, and there are none).

The distinctively Davidsonian catch that we see in the ambiguity of (1) is that under another and peculiarly human descriptive vocabulary, mental descriptions *are* nonphysical, not in the sense that these descriptions are not, say, written in physical ink or spoken in physical words, but in the sense that they do not use any of the words or symbols of physics, and do use other "mental" or psychological words. Mental events can in one sense truly be said to be physical, but they can also be described in a nonphysical vocabulary, just as objects can in one sense truly be said to be physical, but they can also be described in the nonphysical vocabulary of home economics. "That's a cheap bag of tomatoes—let's buy it," we might say. That doesn't mean that the tomatoes are not physical things.

For all its undoubted charm, we should not allow Davidson's view to cause us to forget the logical difficulties with central-state materialism, of the type or token variety, the powerful insights of functionalism, nor the difficulty that any form of materialism has dealing with qualia. When we allude to qualia, to colors for example, we are not just adopting a funny new vocabulary, in addition to talk about electromagnetic radiation, that happens to suit us in our dealings with the world. The new vocabulary is not just a different way of talking about electromagnetic radiation.

It is a way of talking about something completely different: colors. Colors have properties that are not strictly physical—for example, brightness. Brightness is not physical. It is related to *luminance*, the narrowly physical and physically defined concept, which is about how much radiation is transmitted, emitted, or reflected by a particular unit area. When we say that yellow is a bright color, this has an entirely nonphysical meaning, one that can in principle be determined and can only be determined by direct observation, without the measurement of luminance of the yellow colored area. *Brightness* is not a concept to be found in physics but, on the standard view, a concept to be found in psychology.

Eliminativism

With anomalous monism, one has the feeling that the mental has been spirited away, as some, not including Davidson, might think it deserves to be. Perhaps it would be better for Davidson to allow that that there is no such *thing* as the mental, though there are mental *vocabularies*, descriptions, explanations and ways of talking, or mental concepts, but then one starts to worry that the mental is being swept under a convenient linguistic rug.

With the philosophy of mind known as "eliminativism" or "eliminative materialism," we have the straight recognition that talk about the mental will not fit into the scheme of things given to us by the study of the physical world as it applies to human beings, or to any other part of science. Eliminativists admit that mental concepts and terms cannot be reduced to scientific physiological ones. They draw the conclusion that in a completed neuroscience there is no need and no room for mental terms and concepts, and that statements about things mental are just false. These statements are relics of an outmoded psychology and psychophysics, just as statements about witches are relics of outmoded an outmoded view of human nature. There are no witches, and witches are a product of superstition. Nor can we relate witches to concurrent physical events, such as pot-stirring and the training of cats by night. Similarly, says the eliminativist, there are no hopes, fears, beliefs, and desires; they are a product of an inherited form of language that has no basis in science, explains nothing, and has no use beyond the parochial view that belongs in the gossip-filled village shop, and certainly has no use in a scientific laboratory.

Unlike Davidson, eliminative materialists, of whom the most distinguished are Paul and Patricia Churchland and Stephen Stich, take the view that the sentences of the psychology of everyday life that refer to hopes, fears, beliefs, and desires are a sort of a theory, but a completely

Eliminativists admit
that mental concepts
and terms cannot be
reduced to scientific
physiological ones.

false one. What is called "folk psychology" by its detractors, on the analogy with "folk remedies," "folklore," and so on, is false:

> The common-sense conception of psychological phenomena constitutes a radically false theory, a theory so fundamentally defective that both the principles and the ontology of that theory will eventually be displaced, rather than smoothly reduced, by a completed neuroscience.[7]

For Davidson, on the other hand, folk psychology is not explanatory, and it is not a theory at all. That role is reserved for physics. But it is descriptive.

Folk psychology, writes Paul Churchland, "suffers explanatory failures on an epic scale, ... has been stagnant for at least twenty-five centuries, and ... its categories appear (so far) to be incommensurable with or orthogonal to the categories of the background physical science whose long-term claim to explain human behavior seems undeniable."[8] Most philosophers disagree with Churchland that there is something called folk psychology which is a *theory* that makes predictions, the so-called *theory theory*. It is rather a loose set of *concepts* that we employ in ordinary life, concepts like *family*, in the social world, or the *state*, in the political world, or *work of art* in the world of art. Propositions using such concepts are not radically false. And nor should such concepts be swept away in return for some supposedly more profound and accurate scientific concepts, any more than those of cooking or of personal relationships.

There is another problem with Churchland's claim. Arithmetic, for example, has stagnated for far longer than folk psychology, if that means merely that it has not changed. There have been no changes in elementary arithmetic since it was discovered. Multiplication, division, addition, and subtraction—all have "stagnated." Projective geometry, to take another more high-powered example, is in essence complete, and has been since the late nineteenth century. Why then should the concepts of the psychology of ordinary life also not remain undisturbed?

The answer, writes Churchland, is that folk psychology should be displaced because it has not explained mental illness (all of it?), creative imagination, individual differences in intelligence, sleep, the ability to hit targets with projectiles such as baseballs, 3D perception, all the visual illusions, memory, the speed of memory, learning (including learning in prelinguistic infants), and so on and so forth. On all these, folk psychology sheds "negligent light."

The list sets a high bar indeed—too high. For *science* has not explained what sleep is, nor what mental illness is (which incidentally it could not do on the eliminativist view, since "mental illness" is a folk psychological concept, and so it must be "radically false" that people have mental illnesses), nor what creative imagination is (*imagination* is another folk psychological concept, however), and so on. Perhaps there is some conviction that *only* science could explain all these mysterious things, but that is the *conclusion* for which Churchland is arguing, not a *premise* from which he is entitled to argue.

And consider. Elementary arithmetic has failed dismally in the last two thousand five hundred years, and indeed the whole of mathematics has failed, to determine whether there is an odd perfect number, the truth of the Goldbach conjecture, the solution to the Collatz problem, the twin prime conjecture, and so on. Arithmetic has really stagnated, no? And it should pull up its socks. Perhaps we should replace it with neuroscience, which has solved all sorts of important problems.

It is again very obvious how eliminative materialism solves the mind–body problem. Proposition (1), that the mind is a nonphysical thing, is false, not because the mind *is* physical, but because there is no mind. Nothing, including the mind, is nonphysical. The existence of something called "the mind," and all its works, is part of a "radically false" folk mythology.

Again, we have a completely successful solution to the mind–body problem, and again we have a view that is itself every bit as hard to believe as the mind–body problem is said to be intractable. The clear success of a solution seems to stand in inverse relationship to its believability.

ANTIMATERIALISM ABOUT THE MIND

Introduction

At the very end of his fine book *Philosophy of Mind*, published in 2006, the distinguished American philosopher Jaegwon Kim writes that the "limit of physicalism" is qualia. Physicalism can be defended, he thinks, for everything except qualia. Qualia cannot, like everything else mental, such as intention, be functionally defined, Kim thinks, and qualia cannot be reduced to anything physical; nor can they be defined at all. Yet Kim is still a proponent of a naturalistic worldview, a worldview that includes mind. How can this be? He writes in *Physicalism, or Something Near Enough*, that "physicalism is not the truth, but it is the truth near enough, and near enough ought to be good enough."[1] This is stylistically good stuff and a good way to end a book, but it simply will not do from a philosophical point of view.

Over here is a worldview, physicalism, which claims that everything is physical. Over there is a clear case, according to Kim himself, of something nonphysical, with a probably potentially infinite number of instances: all the colors, all the sounds, all the smells, all the tastes, all the objects of the other sensory modalities, and all the objects of sensory modalities that we do not experience, if there are any, for example the ultraviolet perception that bees have, their perception of polarization, and so on. To be fair we must also include all the *nonsensory* "what it is like's," all the shades and mixtures and degrees of anger, for example, or depression, or confusion, or elation, or delight, transport, ecstasy, joy, exhilaration, glee, bliss, and on and on. So we have a theory to which there is an infinitely extensible counterexample, and Kim says that is "near enough." Near enough to *what*, one wonders? Not the truth, most certainly. If we conjoin the truth of physicalism with the truth of the proposition that millions of nonphysical color qualia and all the rest can exist, then what we have, by straight logic, is a falsehood, since the second proposition contradicts the first. The conjunction of a truth and a falsehood is a falsehood. How is that falsehood "near enough" to the truth? It seems to amount to something like "If physicalism were only true, though it isn't, it would be true."

Kim is a philosopher with no phobia about metaphysics, so it is hard to understand why he did not start fresh, saying to himself, "Here is the situation. Everything suggests physicalism; but it is false. For one very important class of irreducible entities stands against it." And *then* he might perhaps have asked the question, "How can *that* be? How on earth can that be how things are? How can it be that everything points one way, but the truth lies in the opposite direction?" Kim's blind spot about this may have to do with the fact that colors and the other qualia are apparently causally inactive. His own work has been devoted to the topic of causation and the application of the concept to a variety of philosophical problems; causal inactivity, I suspect, is for him "near enough" to nonexistence. But this is just prejudice against noncausal concepts.

Next I want to examine some well-known arguments, three in number, all going in roughly the same direction, that have produced what some have regarded as an antimaterialist or antiphysicalist tendency in the philosophy of mind recently. The three arguments that I will consider, in their different ways, record the fact that qualia are indeed a *problem* for physicalism, or worse, that the existence of qualia is a *counterexample* to the claim of physicalism that everything, including the mind, is physical. Proponents of these arguments have sometimes been lumped together by others as *mysterians*, but the label is unhelpful. None of the arguments has as its conclusion the proposition that anything is *mysterious*. Their only conclusion is the very unmysterious proposition that physicalism is false. Before looking at the arguments themselves, I will say something about a view that shares with the three arguments the conclusion that physicalism is false, but has little or no appeal for most people, though it was the dominant philosophy in the religiously tinged philosophical atmosphere of more than a century ago.

Idealism

To be antimaterialist or antiphysicalist about the mind one does not have to accept the larger claim made by idealism. "Idealism" is a name given to a number of different philosophies of mind, prominent in the nineteenth century, and no single account of it has been universally accepted by philosophers. Idealism is a metaphysics that tells us something about the nature of reality, as a metaphysics is supposed to do. Just as physicalism tells us that everything is physical, and materialism tells us that everything is matter, idealism tells us that everything is spiritual, or that everything is mental. But what does this mean? A minimal way of stating the claim is that reality is nonphysical, so that idealism is the contrary of physicalism. At the least idealism is antiphysicalist.

This formulation of idealism has a big advantage. If we take reality to be everything that exists, then if the body exists, idealism asserts that the body is nonphysical. So if as we have seen the mind–body problem is the problem

Just as physicalism tells us that everything is physical, and material- ism tells us that every- thing is matter, idealism tells us that everything is spiritual, or that everything is mental.

of squaring the four propositions in our inconsistent tetrad, idealism easily solves the problem by denying the first proposition, that the body is physical. For according to idealism, *nothing* is physical. So there is no difficulty about nonphysical and physical things interacting, since there are no nonphysical things.

Two big questions remain. The first one is how anyone could believe such a view. How could one believe that the body is nonphysical? In one extremely common English language usage "the body" is taken to *be* the physical part of the human being or the organism, whether or not there exists any part other than the physical part. In this usage it would actually be contradictory to say that the body is nonphysical, since that would be to say that the physical part of the human being, whether or not there exists any part other than the physical part, is nonphysical.

There is also a view called phenomenalism, however, descended from the work of George Berkeley and David Hume, which analyzes statements about bodies, including human bodies, into statements about actual and possible experiences or "ideas," in the terminology of John Locke, Berkeley, Hume, and the other British empiricists. If it were successful, this program of translation would *preserve* the truth of every statement about physical bodies, while understanding them at bottom as statements about possible or actual experiences or sense data. To say that there is a sandwich in front of me is to say that there is a whitish-brown trapezoid in my visual field, with yellowish fringes (that's the cheese hanging out of the edges of the sandwich), and so on, and also to say that the whitish-brown trapezoid will disappear between two pink strips (that's my mouth, phenomenalistically interpreted) in the next ten minutes, and so on.

A number of objections to phenomenalism have carried a lot of weight, such a lot of weight that there are few phenomenalists (or idealists) left. To my mind the biggest objection is that there is no explanation as to why the experiences appear in the sequences they do. Nonphenomenalists will explain this by a very natural reference to physical objects and their behavior. The reason the perceptual trapezoid disappeared between the two pink strips in my field of vision, says the nonphenomenalist, is that the sandwich went into my mouth. But this explanation is not available to the phenomenalists. They will have to start by saying that the trapezoid disappeared between the pink strips because the sandwich went into my mouth, but then for them this second statement comes down to the statement that the trapezoid disappeared between the pink strips. However, a phenomenalist who takes the phenomena to be both sensed and unsensed objects of experience, or what Bertrand Russell called *sensibilia*, can deal with this worry. The forthcoming explanations are just the regular explanations of physics and the other sciences, and of the common sense that goes along with them.

So there are ways to defend phenomenalism at this point, but to my mind a deeper question for idealism is how mind and body interact, given that *neither* is physical. If we are idealists, bodies do not have linear dimensions or a position in space. One still wants to know how minds and bodies interact. There is something very unclear about what idealism asserts, both about mind–body and body–mind causation or interaction. Physical things interact with physical things because physical particles push physical particles along. Does one thought interact with another by mind-particles pushing mind-particles along? But there are no mind-particles. Is it just a matter of magic, then? Behind these difficulties is a mystery about mind–*mind* interaction. How do mental things, such as feelings, interact with other mental things, such as thoughts, since neither of them possesses physical mass and energy to fuel the causing? This last question is of course a question not just for idealism, and so I shall set it aside as we look at the three arguments for antiphysicalism. Though the interaction of the mental with the mental is a fascinating problem, it is not really a part of the mind–body problem. The mind–body problem is not the mind–mind problem, whatever light it may shed on the mind–mind problem.

Three Important Antiphysicalist Arguments

At the end of the twentieth century and the beginning of the next, physicalism seems largely to have run its theoretical course as a solution to the mind–body problem. Although most philosophers are probably physicalists today in spite of this, when toward the end of the last century a number of significant antimaterialist arguments appeared (I discuss one from Thomas Nagel, one from David Chalmers, and one from Frank Jackson), there was no unanimity of response from the physicalists.[2] It would be worthwhile to have a study just of what the physicalist responses to the arguments were and how they worked, what in German is called a *Rezeptionsgeschichte*, a history of the reception of the arguments. The truth is that for physicalists the *Rezeption* seems to have been all over the map. Most of the critical responses *were* physicalist, of course, because typically the antiphysicalists like me were happy to see the arguments prospering philosophically, if generating controversy is what philosophical prospering is.

It is also important to know something about these arguments because they may tell us something about *what* it is that has eluded physicalism, and, just as importantly, they may tell us *why* it has eluded physicalism. The arguments themselves do not offer a solution to the mind–body problem, and they have been combined with a number of different solutions. For example, some have taken Nagel's argument, inaccurately in my view, as a support for simple psychophysical dualism, whereas in reality it is an argument for skepticism about our grip on the mind–body problem, combined with some intriguing hints about how this skepticism can be overcome; Chalmers's argument has been offered in support of both functionalism (though not about qualia) and dualism of all sorts, not just his own "naturalistic dualism"; and so on.

Nagel's argument is perhaps the most dramatic of the three arguments, but it has the logically weakest conclusion of the three. His conclusion is not that physicalism is false, but that though it is true, we do not see how it *could* be true. We do not understand *how* it could be true that our experience is physical, much as someone leaving a chrysalis in a box might not understand how it could turn into a butterfly by morning. A physical explanation is objective, but "the phenomenological features of experience"—qualia—are subjective. They belong to a particular point of view, which is ours, and they cannot be detached from that point of view. The physical line of thought, however, will "gradually abandon" that point of view, leaving us with no understanding at all of the subjective.

Nagel asks us to imagine trying to understand what it is like to be a bat. The subjective experience of a bat, he claims, is one that is closed to us, with our entirely external knowledge of it. Objective phenomena, such as lightning and thunder, can be understood completely and objectively. The subjective experience of the alien character of the bat's consciousness can be understood neither completely nor objectively.

> It will not help to try to imagine that one has webbing on one's arms, which enables one to fly around at dusk and dawn catching insects in one's mouth; that one has very poor vision, and perceives the surrounding world by a system of reflected high-frequency sound signals; and that one spends the day hanging upside down by one's feet in the attic. Insofar as I can imagine this (which is not very far), it tells me what it would be like for *me* to behave as a bat behaves. But that is not the question. I want to know what it is like for a *bat* to be a bat.[3]

It seems to me one could have a vivid hallucinogenic experience, one that turned out to be exactly like the experience of a bat, though one might never know that it was accurate. The hallucination might even include the experience of living in a cave with other bats, an experience that one subsequently discovered to be completely accurate, perhaps by visiting the cave. It is doubtful whether there is any particular limit on what is imaginable, and that includes what is logically impossible. Philosophers are more or less agreed on the imaginability of the logically impossible. Seen from this point of view, Nagel's point really expresses the dependence of imagination on sense experience. His problem is not the so-called other minds problem. That is the problem, like the mind–body problem going back in its most unregenerate form to Descartes, of how we can know the mind of another, in addition to knowing the external condition and behavior of that person's body, presented to us in experience. Nagel's problem is rather the problem of a systematic gap in our knowledge due to a particular biological limitation, a limitation we have as a group or species. We cannot know what it is like for bees to see ultraviolet light, for example, without to some extent—exactly to the extent that we come somehow to possess the ultraviolet perceptual systems of the bee and cease to be ourselves— *becoming* the bee.

However, one cannot argue that I cannot imagine what it is like to be you on the ground that then I would have to *be* you. With those we know well, perhaps especially when they are in trouble, we can imagine without difficulty what it is like to be them. We have a greater empathy than Nagel allows. Indeed, if his argument works, it establishes that we can never have *any* empathy at all, and that we are all psychopaths. If empathy is what it is commonsensically taken to be, which is to be able to feel the feelings that another has, then empathy is logically impossible.

I do agree with Nagel, however, that our experience cannot be reduced to the physical, though not for the reason he gives. The phenomena of sound cannot be reduced to waves, color to electromagnetic radiation, and so on. But this is not because the phenomena express a particular point of view or subjectivity. It is instead a result of the simple fact that colors and sounds are not waves and they possess properties that are incompatible with the properties of waves. Sounds do not have amplitude, for example, though they do have volume, so it makes no sense to ask for the amplitude of a particular sound, rather than the volume. Colors do not have amplitude, though they do have brightness, so it makes no sense to ask for the amplitude of a particular color.

It may help to try to locate Nagel's view on the map of the mind–body problem given to us by the inconsistent tetrad with which we started. Nagel certainly accepts that the body is physical, and that the mind and the body interact. He also sees that physical and nonphysical things cannot interact. He accepts the gulf between the physical and the phenomenological. So he is stuck with the question how it can possibly be true that the mind is physical, which is what he wants to believe anyway. We cannot understand, he thinks, how it *could* be true that the mind is physical, though it is, despite the fact that we can understand and even have evidence that it *is* true. What is left is just that there is a difficulty for physicalism, and Nagel suggests that the path toward what he rather mysteriously calls an "objective phenomenology" is the right one to take.

I suspect that the concepts of *imagination* or *conception* and *possibility* may play very much the same role in the Australian philosopher David Chalmers's famous zombie argument against physicalism as they do in Nagel's argument. In Chalmers's article of 1995, "Facing Up to the Problem of Consciousness," and in his 1996 book *The Conscious Mind*, he argues that our world contains consciousness, but that we can conceive of a world exactly like it physically—*physically* identical—in which the creature corresponding to Chalmers, say, and identical to him physically, does not have consciousness.[4] This creature would be a Chalmers zombie. From the possible existence of this entirely physical creature, it follows that consciousness is not physical—for if it were, the zombie Chalmers would have consciousness in virtue of its physical characteristics, in particular the neurophysiological ones. Just by existing it would be conscious. The zombie argument, by the way, has a history before Chalmers that goes back earlier in twentieth-century philosophy, and can ultimately be traced to Descartes's considerations concerning the possibility that the mind should exist without the body.

There are interesting arguments against the possibility of zombies, but none of them are particularly convincing, to my mind. For example, suppose that Chalmers smells his morning coffee and says, "I smell coffee." What he says is true. But what about the zombie Chalmers? He (or it) does not smell coffee, in the sense that he has the appropriate qualia, so when he says "I smell coffee," in this sense, his statement is false. So, the objection goes, the two beings are not physically identical. But of course there is no real difficulty here, since truth and falsity have never been supposed to be physical concepts, or not by many, rather than semantical ones, or if they have they should not have been. The difference between the truth of what Chalmers says and the falsity of what zombie Chalmers says does not constitute a legitimate *physical* difference. It consists of two logical relationships between what Chalmers says and what his zombie twin says, and the facts.

Much of the argument directed against Chalmers's zombies has been about the *possibility* of zombies, and has deployed sophisticated considerations concerning abstract possibility. *Are* zombies possible? Could they exist? If we say that they can, we seem to be begging the question against physicalism, for we are assuming that the physical zombie is not conscious, and that the physical part of the zombie is not responsible for the zombie's consciousness, since there is no such thing. If on the other hand we say that nonconscious zombies cannot exist, we seem to be begging the question against antiphysicalism, by just assuming physicalism.

A simpler though inconclusive argument against the zombie argument is that saying that the zombie is not conscious begs the question against physicalism. A central-state materialist, for example, will say that the physical part of Chalmers, which includes his brain, *is* conscious, and that to say otherwise in a premise merely *states* but does not argue for the denial of physicalism. The trouble is that it will not do to assume that the zombie *is* conscious, for by the same token that would assume without argument that physicalism is true. It is not clear where the victory lies here, but what Chalmers has to say in later papers about a positive solution to the mind–body problem for qualia seems to be a form of property dualism. Mind and body interact, because they are not distinct, so Chalmers's position is not dualism. But the mind does have nonphysical properties, as shown by the zombie argument. For Chalmers the mental does not reduce to the physical. Chalmers is a property dualist, but with a difference. The difference is his treatment of the proposition that the mind is nonphysical. In one sense Chalmers denies this proposition. The mind is perfectly physical. But in another sense, he accepts the proposition: the mind is also nonphysical, in that claims about the mind do not *reduce* to claims about anything physical. We cannot take a proposition about what I am thinking, say, and *reduce* it to a proposition about the neural circuitry in the brain. Nevertheless, my thinking is the neural circuitry in my brain. This position is reminiscent of the earlier central-state materialist's view that though "gene" does not *mean* "DNA," nevertheless the gene is DNA, and that though "mind" does not *mean* "the relevant part of the central nervous system" (CNS), nevertheless the mind *is* the relevant part of the CNS. Both the central-state materialists and Chalmers detect an ambiguity in the first proposition that the mind is nonphysical, and are able to have their cake and eat it, both affirming and denying the proposition, in the two different senses. In one sense, the property sense, the proposition is true, and in another, the substance or thing sense, the proposition is false.

It is still a troubling question, though, in what way mental states *could* be physical states. This is not a matter of what we say or think, but of the way we are to conceive of my thinking of my grandmother in Italy as a set of neurons firing, or for that matter anything "emerging" out of a set of neurons firing or "supervening" on them. That is Nagel's worry. We cannot imagine following a sequence of events in which the sequence of the events of the neurons firing followed far enough will continuously lead to the event of my thinking of my grandmother in Italy. For this reason, Nagel and those who followed him in a similar line of thought (Chalmers, Frank Jackson, Joseph Levine, and Colin McGinn are the most prominent) have been lumped together as "mysterians," who proclaim the mystery of consciousness. But the word is not really a good fit for Chalmers and Jackson, who would be better described simply as antiphysicalist.

Chalmers has also discussed a form of panpsychism that he calls "panprotopsychism." Panpsychism is the view that fundamental physical objects have mental or conscious states, so that the mental is built into the world alongside the physical from the beginning. It is as though God could not help creating parallel mental states whenever he created the fundamental physical states. "Panprotopsychism" looks like a load of typographical errors, but it is not. It is the view that the fundamental physical objects have "protoconscious" states. These are the precursors of conscious states that, though they are not themselves conscious states, can together cause conscious states to emerge from their combination or collection. Collectively they *are* conscious states, but only collectively. Here it seems to me that Chalmers's view looks like a form of emergentism, or perhaps epiphenomenalism. It has some of the difficulties of those views, and perhaps the extra one of seeming to suggest that somehow the protoconscious states are thought to be in some more primitive sense already conscious. For if they lack consciousness individually it really is difficult to see how a collection of them could have it. (This sort of difficulty is known as "the combination problem.") And Chalmers's view seems to inflate the mind–body problem to cosmic proportions. The relation between mind and body will emerge for every part of the universe (this is the "pan" bit) that has a psychic part.

Chalmers himself does write that he is not sure that the arguments for panpsychism are sound, but he also is not sure that they are not sound. His remarks, read in context, suggest at the least a tremendous sympathy with the arguments for panpsychism, to the extent that he seems to be giving his own view, whereas I do not feel the same rush of excitement in his arguments going in the other direction. His own project seems to be one of working out ways in which physical and nonphysical things can interact, or do something that plays the role of interaction. He offers, for example, the hypothesis that *information* might play the role of the fundamental something that has both physical states and states that carry qualia. So information could manifest itself in one way or the other, and this might be regarded as interaction of a sort. It is an interesting speculation, but no more, I think, because it is very hard to see how a sequence of bits in a bitstream—traveling optical information, for example in a telecommunications network, made up of a flicker of successive light and dark states at a point in the fiber, or 0s and 1s—could turn itself into a stream of qualia or consciousness. It would be an event of biblical proportions.

Chalmers is prepared to concede, however, that his is a very speculative theory, though it may just do work to mitigate the epiphenomenal implications of the zombie argument. The zombies behave physically like their conscious counterparts, but in that case there appears to be a problem in understanding how consciousness has any effect on the physical. Chalmers thinks as a result that we must work toward seeing how consciousness and the physical can work together, or in what way it can be false that physical and nonphysical things cannot interact or do something as good as interact. I am not even a little bit convinced by all this, because I just cannot see how something digital like "information" (1s and 0s) could turn itself into the color red. This really is just nonsense. The information whizzing around in a CPU does not out of itself produce red. The color arises on the computer monitor, not from information as such, but from codes yielding coordinated physical and optical effects in the phosphor dots, plus the contribution of the eye, for example in the optical fusion of red and green to produce yellow. (Yellow is not physically present on a TV screen, as can be verified by examining it with a good magnifying glass.) *Which* effects occur is dependent on the information presented to the monitor, but the color that appears does so for the usual physical and psychological reasons detailed in the science of color, principally from the explanation of how the electron beams striking the phosphor dots produce different colors, not from pure information theory.

An argument related to the zombie argument was proposed by Frank Jackson, another Australian philosopher. (Why, I wonder, have two-thirds of the best arguments against physicalism come from Australians? Perhaps it is because the previous generation of philosophers in Australia had more leading physicalists than anywhere else, apart perhaps from the United States, so that the antiphysicalist arguments were a reaction to the prevailing view. Or perhaps it is something in the beer.)

Of course, there are more than three arguments against physicalism, but the three I discuss here have been among the more influential and the most discussed. Jackson's argument in "Epiphenomenal Qualia" is simplicity itself. He asks us to imagine a brilliant color scientist, whom he calls Mary. Mary is brought up and lives in a black-and-white environment. She is "brilliant" in the sense that she possesses all the information given by a completed physical science of color, including neuroscience. She has all of this information at her fingertips. Now comes the day when she opens the door and leaves her achromatic environment. She steps into the fully colored world. It seems that she will acquire some new information, assuming that her color vision system starts to work fairly quickly; she learns something new. Perhaps we might wish to say, though Jackson does not, that she finally learns what red *is*, what blue *is*, and so on. In any case, she learns what these colors *look* like. But if she has learned something new, and gained information in addition to the totality of physical information, then not all information is physical information.

In a separate argument in the same paper Jackson also describes a character called Fred, who sees a color that "standard human observers" do not see. All the physical information in the world will not help an observer (I shall call him F-red) who does *not* see this new color—perhaps it is a shade of red—to know what it is that Fred is seeing. F-red is like Mary before she leaves her room. F-red has all the physical information that there is or could be, but he still does not know what Fred sees. (Jackson is of course assuming that there *could* be a novel color, which is naturally something that has been argued about.)

Both of Jackson's arguments look valid, and if they are, they establish that the mind is nonphysical, or at least that we have nonphysical information about the mind. Sometime after publishing his argument, however, however, Jackson took it back. He had decided that it led to dualism and that the dualism it led to is epiphenomenalist. This meant that the qualitative states whose nonphysical character he had championed, though they exist, are without effect on human behavior. There they are, but they have no effects. Epiphenomenalism is hard to believe, however, not least because, as we saw earlier, it has all the problems of dualism, and more of its own. (*Two*-way epiphenomenalism might be better, but that is just dualist interactionism.) As part of his self-apostasy Jackson came to believe that sensory experience in general is *representation*, so that what is important about it is the information it gives, not its qualitative character. Or rather, its qualitative character *is* representation.

Is this a good line to take? Suppose that there existed a solipsistic two-entity universe, a world with only two things in it: a perceiver, and the perceiver's qualitative experience, a single quale. There is *ex hypothesi* nothing for the quale to represent, but it seems undeniable that the perceiver experiences it. We might think that he could come to enjoy it.

One is bound to feel that the three antiphysicalist arguments have something in common. Their conclusions are not exactly the same, of course. Nagel's argument has the conclusion that we cannot see how our first proposition ("The mind is nonphysical") can be false. Yet it is, because physicalism is true.

The zombie argument starts with the fact that the Chalmers zombie is possible. If the zombie exists, he or perhaps "it" is not conscious. It is physically identical with the whole of Chalmers's physicality. But there is more to Chalmers; Chalmers is conscious. So consciousness is not physical. Like Jackson's argument, Chalmers's argument has as its conclusion the proposition that the mind is nonphysical and that physicalism is false. Jackson realized quickly the epiphenomenalist implications of his argument for the mind–body problem, and abandoned it. Chalmers took the heroic line of trying to see ways in which dualism could be true, and that is what has led him to consider panpsychism. His views come from someone who is no matter what prepared to take the mind–body very seriously, and panpsychism, though it may seem bizarre, is a reflection of this seriousness.

The mind–body problem is not going to be made to go away easily, and neither is consciousness. Consciousness may be less important in the general economy of the mind than the more enthusiastic of the "qualia freaks" suppose, but the three antimaterialist arguments attest to its reality and importance. For Nagel, the mind is physical, but we cannot see how that is possible, and that is the power of the mind–body problem. If we cannot see how something is possible, we are bound to respect the view of those who believe that it is possible. But this is not a solution to the mind–body problem. It is a declaration that physicalism is true, but incomprehensible. The second half of this claim is true, though, even if the first is false.

For Jackson and Chalmers, the qualitative part of the mind is nonphysical, and so they are dualists. Jackson's dualism is epiphenomenalist, and he found that in the end it was not a position he could live with.

Chalmers has stuck to his guns, and he has toyed with exotic theories such as panpsychism that build dualism into the fabric of things. The difficulty here is that he is not giving us an account of the very thing of which Descartes could not give us an account. Even the panpsychist ought to be able to give a coherent account of the relation between the mental and the physical, and he does not. His position is sound enough, to the extent that it does recognize the "data" (the inconsistent tetrad) that fuel the original problem. The trouble is that the mind–body problem

The mind–body problem is not going to be made to go away easily, and neither is consciousness.

metastasizes into the same problem across the entire universe. Why give us infinitely many more instances of the mind–body problem than we already have? It is of course open to the panpsychist to retort that if one instance of the problem is solved, they all are, so the numbers do not matter. Either way, though, it is better to look for a solution where one is to be found.

Let us look next at the scientific solutions that have been offered to the problem.

SCIENCE AND THE MIND–BODY PROBLEM: CONSCIOUSNESS

Introduction

It would be impossible to consider all aspects of the scientific study of mind and mentality as it has developed, even very recently, such as work on memory, attention, and so on, and so I shall take as the representative of the scientific study of mind the recent work that has been done on just of one part of the mind: consciousness. I think that it will give us a fair idea of the way scientific thinking on mind–body relations has been going, although there are undoubtedly important differences between the scientific study of consciousness and the scientific study of the other topics relevant to the mind and the mind–body problem.

The 1990s saw a surge of interest among scientists and philosophers in the topic of consciousness, after it had languished for about a hundred years. This is not a book about the history of ideas, but it is reasonable to speculate that part of the explanation for the reappearance of interest in consciousness may have been that the existing materialist or physicalist philosophical theories had been played out and found wanting. Consciousness had resisted the depredations of reductionist philosophical theories, and so it became acceptable to start to think freely about it again, from a scientific point of view. What else was there to do? Philosophers, of course, or some of them, had never stopped thinking about it. The paradoxical result has been a resurgence of interesting and powerful materialist or physicalist theories from the scientists, as they found themselves actually able to come to grips with the elusive, invisible, and shapeless beast. The work of Giulio Tononi, for example, proceeds in part by attempting to elucidate the concept of consciousness, just as the philosophers had always done, rather than taking it as an indefinable datum and trying empirically to line up some neural correlates for it.

Descartes's view was that the mind, thinking, and the self are all the same thing, and it is a nonphysical one: a conscious one. That view allowed him a sweeping simplification and an extraordinarily penetrating view of the status of mind in its connection with knowledge. But it is clearly wrong: there are "parts" of the self that are not conscious; some thinking goes on without the self; the role of "the self" in consciousness is in any case unclear; and so on.

There are "parts" of the self that are not conscious; some thinking goes on without the self; the role of "the self" in consciousness is in any case unclear.

It is worth looking at the scientific accounts of consciousness, if only because even at this late date they still do not on the whole seem sensitive enough to the issues that allowed the philosophers first to discover and then to work on the mind–body problem.

Baars and the Global Workspace Theory

As early as the 1980s Bernard Baars had begun to propose the "global workplace theory" of consciousness, and he has written about his own theory extensively since then, both by himself and with others. The theory remains a completely physicalist one, as I see it, in spite of the suggestive Cartesian imagery that Baars is eager to use. He writes, for example, about a conscious "inner theater," onto which, or better *into* which, information derived from the senses is "projected." The phrase "inner theater" derives from Gilbert Ryle's *The Concept of Mind*, and it was intended by Ryle to be part of a mocking characterization of Descartes's conception of the mind and its relations to the body. Ryle had also described "the Cartesian Myth," with equal distaste, as "the Myth of the Ghost in the Machine." The reason Baars is relaxed about the dualist *imagery*, including the description of consciousness as a spotlight in a dark theater, is that he actually conceives the global workspace entirely in terms of the physical activity of neuronal cells.

There is, he claims, a network of cells in the brain, which he calls "the global workspace." It is rather like a working blackboard onto which various images and statements are "projected." (This of course raises the question of the relationship between the images and statements and the firing of the neurons, which is the mind–body problem!) Some of the images and statements are erased. But some survive. If cells from the separate regions of the brain, such as the regions dealing with visual consciousness, auditory consciousness, and so on, all of which are more or less localized, provide signals to the global workspace, and these signals are chosen and broadcasted by the decentralized network, then Baars claims we have consciousness of the information that is sent out.

It is a defect of Baars's view that what we are conscious of is neither things nor qualia, but information. It is however also not clear whether it is the mere presence of information in the global workspace that constitutes, or provokes consciousness, or whether it is the broadcasting itself that is the consciousness.

Baars strenuously denies the relevance of his theory to the "hard problem" of consciousness, but then it is difficult to see then how it could be a theory of *consciousness* at all. Baars seems in fact to favor a kind of emergentism in which consciousness *emerges* from the global workspace, when strictly what his theory entails is the identity theory:

consciousness *is* the global workspace at the time of the broadcast.

Another puzzling aspect of the theory is that the incoming sensory signals do not have to end up in a single spatial location, as the global workspace is distributed throughout the brain. One might wonder why the presence of information in different locations within the global network would count as integrated enough to be consciousness. Still, where the active cells are to be found is presumably an empirical question, and it is also an equally empirical question in psychophysics whether activity in the global workspace is indeed associated with consciousness. Recently, after a quarter of a century, researchers have found a certain amount of what may be evidence for Baars's theory. In 2009, recordings from electrodes already in place in the brains of patients suffering from epilepsy showed an increased coordinated activity over a large part of the cerebral cortex, which has especially dense connections, during conscious perception. On the other hand, this evidence could also be evidence for several other theories, among them the ones discussed later on in this chapter.

One of the interesting things about Baars's theory is what it has to say about the *function* of consciousness. What consciousness does, according the global workspace theory, is to integrate and select information, linking large numbers of different neural networks, and to make the integrated information available for decision making, action, deliberation, imagination, and so on. The important thing that the theory has to offer is the proposal that what is placed in the global workspace in short-term memory is subsequently available, though not for long (for a time on the order of 100 milliseconds), to these other psychological processes. The later processes take the material of consciousness, the output data from the workspace, as their material, and use it in their own characteristic ways.

> This is the primary functional role of consciousness: to allow a theater architecture to operate in the brain, in order to integrate, provide access, and coordinate the functioning of very large numbers of specialized networks that otherwise operate autonomously.[1]

From this point of view, however, it looks as though what Baars is proposing is also functionalism. The concept of the global workspace is a not the concept of a physical place in the brain, but of functional relationships between what is happening in the various networks. What makes his theory so hard to follow is that it is a mixture of emergentism, the identity theory, and functionalism, and Baars shows no awareness at all that his view mixes up incompatible philosophical theories of mind.

Crick and Koch, the 35–70 Hz Hypothesis, and the Claustra

The mind is not, however, particularly unified, in the sense that, *pace* the global workspace theory, one thing goes on in it perfectly independently of another. A stream of disconnected thoughts, some unconscious, can be interrupted by a memory, or the results of an unconscious calculation can suddenly present themselves for no apparent reason. "Oh! So that's the answer!" we say. The unity of the mind, whatever exactly it might be taken to mean, is partial and relative to a time and to a place, and to the previous state of the mind, and no doubt to other things as well. I take all these to be empirical facts.

Many have also taken it to be an empirical fact, something observable, in some way, that the same thing is *not* true of consciousness. Consciousness is *unified*. It presents itself as one "field" of experience; or at least so it is thought. I am skeptical of this claim, and the metaphor of the "field" adds little, or nothing, or less than nothing, to it.

One has only to think of the fading of visual awareness at the edges of what one can see to realize that even visual consciousness is not a monolith. Its intensity fades with age, it has holes in it, such as the blind spot, and its power can vary depending on mood. There are also phenomena such as visual agnosia, the inability to recognize objects, that interfere with it. And there is sleep, which knocks great holes in the temporal unity of consciousness.

There is a strong linguistic element to the claim that consciousness "presents" itself as any sort of unified "field"—what, after all, would a disunified *field* be? The linguistic element here is that "consciousness" is not only a noun but also an adverb when followed by a preposition: "I was conscious of a large cow behind the hedge." Or it can be followed by a relative pronoun introducing a noun clause, such as "that": "I became conscious that I had been deceived." So the objects of consciousness are bound in good grammar to be there, delivered to the subject, in consciousness, unless the claim to consciousness is false. It is rather as if one had a concept, *anything that's in the bag*, and then, looking at all the things in the bag, made the surprising discovery of the "field" of the bag, or the "unity of all the things in the bag," or "the unity of bagness." If something is not in the bag, then it is not part of this unity. And if the elements of consciousness were not conscious, then they would lie outside consciousness, and not be integrated into the unified field. So the unity of consciousness seems to be just an artifact of the grammar of "conscious."

There is another obvious problem here. The bag has a material, say, sackcloth, which literally contains the things in the bag. But consciousness has no material. It needs none, for there is no danger that the "elements" of consciousness will fall out of consciousness if their containing material ceases to contain them—for then they would simply not *be* elements of consciousness.

The linguistic type of view I have been giving here is not at all a popular one today, though it might well have found some favor fifty years ago. Some cognitive scientists have tried to pin down what they take to be the phenomenon of the unity of consciousness by contracting it into a narrower and apparently more tractable problem known as "the binding problem." Suppose I am looking at a red square and a blue circle. Somehow, according to these cognitive scientists, I end up seeing what is actually there (a red square and a blue circle), and not a red circle and a blue square; I see all the elements organized in the right way, rather than as a jumbled collection.

Then there is the larger "intermodal" binding problem. Visual elements find themselves within consciousness, but so do auditory elements, tactile elements, and all the rest. The deliverances of all the different sense and of all the contents of our minds come together in consciousness. What brings all these different things together into one unity?

The difficulty addressed by both nonintermodal and intermodal binding is that the areas of the brain that deal with, say, color and shape, or color and sound, are completely different areas. Color information is processed in the brain areas V1, the primary visual cortex, V2 and V4, whereas shape is dealt with in LOC, the lateral occipital cortex, which deals with object processing or shape. Indeed, there is evidence from fMRI (functional magnetic resonance imaging) suggesting that V1 activity is actually *suppressed* by LOC activity.[2] Color information and shape information are separated in the brain. They presumably must be brought back together. Where in the brain does this occur? And how are they brought together in the right way, so that the red square and the blue circle are not turned into a blue square and a red circle?

It may be, of course, that the answer is that they are not brought back together anywhere, or by any means, in the brain. This would certainly interest the philosophers, but it is fair to say that it would be a disappointment to the scientists, and it would leave unanswered the question of why it is that we see a red square and a blue circle rather than a blue square and a red circle. On the other hand, it is difficult to see what help it would be to bring all the information together in one physical place in the brain.

It would certainly be a wonderful thing to find the answers to such questions at a stroke, empirically and scientifically, since (1) inspired guesswork is almost certain to be wrong, and (2) the mechanism by which the red square and the blue circle are put back together in the human brain is not the sort of problem that will yield to guesswork rather than careful physiological work and cognitive science. They are the sort of questions to which Francis Crick devoted the later part of his life at the Salk Institute, working with Christof Koch, in pursuit of a mechanistic or physicalist understanding of consciousness based on neuronal activity. However, they gave not one but two answers, answers that are actually inconsistent.

The earlier hypothesis that Crick and Koch advanced appeared in a paper published in 1990.[3] "At any one moment," they wrote, "some active neuronal processes correlate with consciousness, while others do not. What is the difference between them?" Here Crick and Koch *distinguish* between "neuronal processes" and consciousness, which seems to make them dualists. And in *The Astonishing Hypothesis*, Crick wrote that "our minds—the behavior of our brains—can be explained by the interactions of nerve cells (and other cells) and the molecules associated with them."[4] So on the one hand we have the "behavior of our brains" and on the other the "our minds," which can be "explained" by the interactions of the nerve cells. This sounds like epiphenomenalism. The impression is reinforced when we realize that Crick and Koch are looking for the "neural correlates" of consciousness, or "NCC," as they are sometimes called. For if x and y are correlated, then $x \neq y$. One thing cannot be *correlated* with itself. It is *identical* to, or with, itself. It takes two to correlate.

In their 1990 paper, Crick and Koch endorsed the hypothesis that what differentiated the cells that make consciousness from those that do not is the frequency at which the consciousness-making cells fire—between 35

Hz and 70 Hz, or "40 Hz" for short. When the cells fire synchronously, or alter their membrane voltage (the difference in electric potential between the inside and the outside of the cell) all together, as one, behold: consciousness! "It seems likely that, for one reason or another, certain neurons in the cortical areas involved tend to oscillate at around 40 Hz." By 1998, in an article written as a "Commentary" piece for the journal *Nature*, Crick and Koch were prepared to concede that the idea was wrong: "We no longer think that synchronized firing, such as the so-called 40 Hz oscillations, is a sufficient condition for the NCC [the neural correlates of consciousness]."[5] Their new approach was to imagine cells or the cortical neural networks that run on them to form temporary groups or coalitions of functionality, and it is these coalitions, when they form, that are the basis of consciousness. This seems a healthy development in its way. Synchronization, after all, by itself, does nothing. We can imagine the instruments of the orchestra all playing in time, to the beat, and synchronized to the conductor. But this will not tend to "integrate" the music even slightly if the members of the orchestra are *spatially* separated. If the violins are in Sydney and the cellos are in Reykjavik, there is no binding. The instruments must *also* be spatially integrated for the "information" to be bound. And even if there is a general spatial binding, one can imagine all the telephones in New York City ringing at the same time with the same beat or frequency and volume. Would this have the slightest tendency to produce anything resembling consciousness or a higher-level phenomenon? It seems clear that it would not integrate the ringings at all, though a lot of people would separately get annoyed.

Another coordination problem remains once the 40Hz hypothesis has been abandoned, namely, how are the temporary coalitions of neurons able to synchronize *their* activity? Here, in Crick's last published paper, also coauthored with Christoph Koch, which appeared in 2005, after Crick's death, he and Koch explored the physiology and anatomy and functionality of the *claustrum*, a thin layer of neurons below the neocortex. The title of their paper was "What Is the Function of the Claustrum?" Their answer is metaphorical, as they themselves admit. It is the analogy

> of a conductor coordinating a group of players in the orchestra, the various cortical regions. Without the conductor, the players can still play but they fall increasingly out of synchrony with each other. The result is a cacophony of sounds.[6]

Evidence in favor of such a hypothesis is the extraordinary connectivity of the claustrum, and Crick and Koch argue for an experimental investigation into its role in consciousness. There is no doubt that such an investigation might advance our knowledge of the role of the claustrum, but would it advance our understanding of consciousness, and of the mind–body problem? Crick and Koch's proposal runs headlong into the problem that all interactionist dualists face, when they offer some physical structure or system (say, V1 and V2 for visual consciousness), or the pineal gland, or the claustrum, as the locus of interaction: Why should *this* particular structure or set of neurons be the site of the interaction? Or how can the claustrum activate the mind if proposition (1) is true and the mind is nonphysical? Crick and Koch's view is actually an inconsistent and unstable mixture of central-state materialism and an interactionist dualism with an emergentist slant. The question they do not answer is how the claustrum can project its activity into the mind, or how the activity of these neurons can have an effect on consciousness. There is no solution to the mind–body problem here, and we are back with Descartes. The claustrum may be a better bet than the pineal gland, but we still want to know how the neurons, no matter how well coordinated, could produce mind and consciousness. It seems to me that Crick and Koch did not have the measure of the true difficulty of the problem, and the kind of problem it is: the logical part of it *must* be solved before the scientific and psychological elements of a solution can begin to have any traction, though they may be true and interesting in their own right. It may be that whenever there is consciousness there are coalitions of neurons contributing their firing sequences in a

The logical part of [the mind–body problem] *must* be solved before the scientific and psychological elements of a solution can begin to have any traction.

way orchestrated by the claustrum. That may be true, but it does not by itself give even a hint of a solution to the mind–body problem, any more than the fact that the pineal gland is light in weight and its parts easily moved gave Descartes a datum with which to address the problem. It is not as though consciousness could lift light things, but had trouble with heavy things.

Tononi and Integrated Information Theory

Shortly after the appearance of Crick and Koch's final paper on consciousness in 2005, Giorgio Tononi, an Italian psychiatrist who subsequently collaborated with Koch, produced a theory of consciousness in what he described as a "manifesto."[7] Tononi's philosophy of consciousness, which grew out of earlier work with Gerald Edelman, is straightforwardly functionalist. Tononi claims that there are processes in the cerebral cortex that integrate information, and that the information they integrate *is* consciousness. He is careless with the way he words his theory, on occasion referring, as many scientists and philosophers do, to NCC or the neural correlates of consciousness, as though by itself this phrase did some useful work. The trouble here is again that the "correlates" of something are not the same thing as the things with which they correlate. Radar blips are the correlates of airplanes or ships, but this means that there are two sets of things, not one: radar blips, and airplanes or ships. Tononi also refers to a "correspondence" between integrated information and consciousness. This raises a doubt as to whether Tononi's view is that the *brain processes* which integrate information *are* consciousness, or whether it is the *integrated information* itself that is consciousness, or whether, whatever it is, integrated information only *corresponds* to consciousness.

Tononi's statement in his 2008 paper "Consciousness as Integrated Information" could not be clearer, however, though it is inconsistent with what he states elsewhere. He writes, "Consciousness is integrated information." That makes Tononi a functionalist, rather than a physicalist, for the amount and the degree of the integration of the information are not strictly physical properties, to be studied by physicists; they are computational or information-processing properties. The same information could be integrated in the same way by all sorts of different physical systems.

Tononi asks us to imagine a camera with a large number of pixels whose information is unintegrated. The information at every point from the imaged scene is independent of the information at every other point. It is atomic. The camera is certainly not conscious. Furthermore, a single photodiode representing the presence or absence of light is not conscious of the presence of light, says Tononi, because it is not registering the absence of the alternatives, such as color or shape. It does not represent light *as* light *rather than* color or shape. This clever idea puts Tononi in a position to say why he thinks that the single photodiode receiving a single kind of information is not conscious.

What if, he asks, every pixel were able to register information in multiple possible dimensions, unlike the diode, and able to pass information continuously to every other pixel, so that the state of each pixel were to become responsive to the state of every other? With this kind of reception of multidimensional information and its integration, Tononi thinks, we are approaching the conscious state, or perhaps we even have the conscious state.

Though it is perhaps true that consciousness is more or less integrated, in the sense that it is not completely disjointed, the converse does not follow. It does not follow that if something is integrated, and this includes information, it has to manifest consciousness. It is hard to see why information, or bits—1s and 0s—flying around in whatever form should be the same thing as consciousness. The obvious thought experiment is to construct a machine with as much information in its various "receptive" parts as you please, which yet remains a machine, in the sense that it lacks consciousness. Is such a thing possible? If so, and it seems to be so, it is hard to see what the link is between large amounts of integrated information, by whatever measure, and the kind of awareness that human beings possess. The measure of the integration of information that Tononi uses is represented by the symbol "Φ," and he brings considerable mathematical sophistication to connecting Φ with the standard concepts of information theory and processing.

Φ by itself does not introduce the manifest richness of consciousness, especially the perceptual consciousness that is Tononi's main area of interest. One can imagine a very highly integrated system of information pathways integrating rather little information or input. So the theory of integrated information introduces a second element of the measure for consciousness, represented by the letter "Q" for "quality space" or *quale*. Quality spaces have received an enormous amount of technical attention since the beginning of psychophysics during the nineteenth century, and they have been developed as tools for scaling in all the separate sensory modalities. Color space, for example, is a three-dimensional array whose dimensions are hue, saturation, and brightness—all psychological concepts. The space is constructed by positioning just noticeably different samples of color next to one another. The geometry of the result is in some ways surprising: an irregularly shaped double cone, with maximum brightness at the top of the cone, darkness at the bottom, and at the off-set "equator" a circuit of hues. Having a high Q-value means that picking a point or an area in the space excludes a huge amount of other possible positions and other information, and so reduces options (or "uncertainty," as

Tononi calls it) by a very great amount. For the standard observer there are by some estimates as many as eleven million just noticeably different points in color space. What this means is that Q has a very high value for a point position in color space, and conveys a lot of information. Q captures the formal complexity of the space. But here traditional philosophical worries about formal relations will intrude. We can imagine a space with exactly the same structure as color space that is isomorphic, given the structure of the honeybee's polarization receptors, to a space (a Poincaré sphere) representing the structure of the honeybee's experience, if any, of the qualia of polarization. Honeybees are highly sensitive to the angle of polarized light, and they use it to navigate very successfully even on very cloudy days. Suppose (which is not true) that the polarization and color spaces are isomorphic. Then Q will not deliver qualitative experience to the honeybee, but simply a formal measure of the geometry of the quality space. And the possession of this geometry by a structure is not at all the same thing as possession of the conscious experience itself.

Now we are to imagine that Q represents a measure for *all* the qualitative spaces of the different human sensory modalities: touch, taste, smell, color, shape, and so on, including all sorts of dedicated cells for things such as facial recognition. Though what we get is a very interesting and very complex geometrical structure that *represents* consciousness, it is still not the same thing *as* consciousness. Tononi has substituted a structural *analogue* of consciousness for consciousness itself. It is as though, in selling us a large building, Tononi were to give us an architect's elevation and floor plans and take himself to have sold us the building itself.

However, Tononi does develop a set of essentially mathematical concepts for describing consciousness, and this is an achievement. The mathematical description begins with axiom-like propositions about the structure of consciousness: (i) consciousness should have different parts, so that it is not an undifferentiated mass; (ii) the information presented should be genuine information about the world; and (iii) it should rule out as many options as possible. That is, it should be "exclusionary." Light versus dark information does not produce consciousness, but achromatic light versus dark versus green versus yellow versus square versus round … approaches something like the richness of consciousness. Furthermore, (iv) the differentiated information should exist in a single field; it should not be disjoint, as it presumably is in subjects whose two cerebral hemispheres have been severed from one another.

Perhaps it took a working psychiatrist with skills such as Tononi's to do the hard thinking about representing the phenomenology of consciousness mathematically. Tononi's work deserves even more appreciation when we understand the way in which the concept of integrated information can be mapped onto the brain functions associated with consciousness. He points out that the cerebral cortex has a high Φ, and is differentiated according to the different sensory modalities. Damage to a large part of the cortex will be associated with loss of consciousness and a vegetative state, or with a loss of the specific kind of consciousness associated with specific areas in the cortex.

Michael Graziano and the Attention Schema

A few years after the appearance of Crick and Koch's neuron-based scientific theory of the neural correlates of consciousness, and Tononi's sophisticated mathematical theory of integrated information, a wider theory emerged, equally scientific, though in a different way. It appealed to social, perceptual, evolutionary biological, and information-processing considerations rather than just to the behavior of the neurons. This may seem, to a traditional intellectual sensibility rooted in nineteenth-century science, an unhealthy mixture, a witch's brew of vague cognitive scientific speculations. In fact, the theory of Michael Graziano and his colleagues at Princeton is a representative example of the kind of thing that goes on in the cognitive sciences today, and it is fully consistent with the claim that consciousness appears where there is neurophysiological computation. The theory looks at the organization of what the neurons are doing into systems with different functions.

Graziano calls the central construct of his theory the "attention schema."[8] If it offers any promise of success, that is not just because of the outline of the information-processing structure of the theory, or of the evolutionary history with which it connects, but also because of the deft philosophical conjuring trick that turns the information-processing structure inside out and, as Graziano and his colleagues see it, offers a philosophical escape from the mind–body problem. Let us first consider what the theory states, and then go on to consider what sort of philosophy it is and whether the philosophy is a success.

First of all, Graziano thinks that the hard problem of consciousness is easy, and that the easy problem is very hard. The suggestion is an interesting one. What he has in mind is that we say that we have consciousness, but that is *all* that we can say about it. All we can say is that we say we have it. Qualia are ineffable, in the traditional terminology. What is it like to be conscious? Well, it is like what it is to be like *that* (there follows and inner pointing) since *that* is what consciousness is. The only property of qualia and states of consciousness is *that* they are, one could say. On Graziano's view, that is *easy* to explain, though the details are hard. It is easy in principle to explain why and when we say of ourselves "I *have* consciousness in my head." The computation involved in this sort of attribution is more limited, rather than more complex, than the computation involved in the easy problems of consciousness. It is a very simple thing to say, "I have this."

Along with birds and other animals, we have attention. We can turn our attention to one thing and away from another. Attention here is not a sort of stream of consciousness squirting out of the eyes, and it is not moving one's eyes from one spot to another, although it may involve looking in a different direction. To attend to *this* rather than *that* is to suppress information about *that* and leave only information about *this*. To perform this information-processing trick we have to be able to control the information and keep track of it all. We do this by making a simplified *model* or mental *picture* or *schema* of our attention, like a general keeping track of his armies with a set of model soldiers and little metal tanks on a table. This is more useful than trying to keep track of *all* the information. It is a sort of summary or simplification of what is going on.

Consciousness is a simplified model or schema of the outline of the activities of our attention filters. And the same computational techniques that allow us to construct a model of ourselves enable us to construct what is called a "theory of mind" for others.

We say proudly of ourselves, just as we do of others, "I (or he) has consciousness," but we cannot express anything more about it. Graziano recounts an amusing story told to him by a psychiatrist friend about a patient suffering from the belief that he had a squirrel in his head. When he was asked such things as how big it was, whether it would show up on an MRI, and how much it needed to eat, the patient would deflect the question. But he continued to insist that though he could say nothing else about it, he had a squirrel in his head. This, says Graziano, is like our own situation with respect to the thing we think is inside our heads: consciousness. We can say *that* it is there, but nothing more.

Graziano also points out that it is remarkably easy to attribute consciousness to puppets and ventriloquist's dummies. They really do *seem* to have consciousness, or more personality perhaps, than they should. We are prepared to say, "They are conscious," without the slightest idea of what the consciousness that we attribute to them is.

It is unclear whether Graziano's claim is that the presence of the attention schema itself *is* consciousness, or whether it is that the presence of the attention schema *plus* the computation or "realization" that it is present, which is consciousness. Graziano says both things, sometimes almost simultaneously, but they are genuinely inconsistent. In either case, however, it seems reasonable to suppose that we could build a machine with the structure that Graziano gives for consciousness. Such a machine would filter information, and thereby have attention in the informational sense. It would then create and update a simplified model representing the state and structure of its attention, it would deploy this schema to determine the presence of "other minds," and it would signal to itself, and to others, the presence of the schema. But it might not have consciousness. Whether it did or not would seem to depend on how things are for the machine, not on the mere presence of the psychological and physiological information-processing sort that Graziano offers.

Concluding Observations

Of the four scientific theories I have considered so far (the global workspace theory, the hypothesis of Crick and Koch, Tononi's account of integrated information, and Graziano's attention schema), none has much independent philosophical interest, in the sense that none introduces a new philosophical theory.

Baars's global workspace theory is functionalism, since anything that plays the role of the global workspace and does the work of consciousness would presumably by his own argument *be* consciousness. Or perhaps it is emergentism. Or perhaps it is central-state materialism. It is hard to say. The "astonishing hypothesis" of Francis Crick is a straightforward form of the identity theory, or central-state materialism. In Tononi's work, consciousness is the integration of information, another functionalist conception. Graziano's theory of the attention schema is also form of functionalism, but, inconsistently, also a form of eliminativism.

It is surprising how easily these theories either fit into the existing philosophical theories of mind, and so have little to offer to the solution of the mind–body problem, or are confused or unclear, or both, just at the very point at which it is most important to be neither confused nor unclear. The mind–body problem is indeed a philosophical and logical problem, as is clear if we accept its formulation as an inconsistent tetrad in Campbell's form. The details of its solution are important, and it will not do to offer even slightly inconsistent philosophical stories.

There are other significant theories of consciousness that I have not considered, not because they are not interesting or important, but because they mainly repeat this pattern. They reproduce in fairly straightforward if hybrid and inconsistent ways existing philosophical approaches to consciousness. Dennett's "multiple drafts theory" of consciousness, for example, looks like a combination of a form of behaviorism with eliminativism. Dennett denies the existence of any qualia at all, and his multiple drafts theory of consciousness seems to introduce a fictional and irrealist element. The biological and logical theory of Roger Penrose and Stuart Hameroff is a form of emergentism. When there is an "objective reduction" of the quantum mechanical wave, one that can take place only under special biological conditions in the brain, the result is consciousness, and it is propagated through the brain with a 40 Hz cycle. John Searle's biological mysticism takes consciousness to be a biological something that is *caused* by processes in the brain, and he very reasonably calls his view "biological naturalism" rather than mysticism, though "biological materialism" would do just as well. Only the brain has the causal abilities capable of producing consciousness. But why should we not reproduce these abilities in silicon?

And though there are other scientific theories of consciousness worthy of consideration, they also do not break new philosophical ground, at least in the area of the mind–body problem, but rely on old philosophical theories. It is instructive to see some of the philosophers mixing it up with the scientists on their own ground and offering what are undeniably testable scientific theories, even if the reverse phenomenon has not been very edifying.

NEUTRAL THEORIES OF MIND AND BODY

Qualia or Phenomenal Properties Again

In these final two chapters, I want to introduce "neutral" theories about mind and body. The neutral theories do not try to extract mind from matter, by tortuous logical means, as physicalism does, or to dissolve matter into mind, improbably, as idealism does. As a result, in theory at least, the neutral theories ought to have less of a problem with the relation of mind and body. Some of the neutral theories start with qualia or phenomenal properties as a sort of given, so it will be good to start by looking again at phenomenal properties. Phenomenal properties are something of a touchstone of any theory. If the theory has a successful account of phenomenal properties, it at least has a chance of being true. If it has none, it is doomed from the start.

It seems obvious that we do experience phenomenal properties. How could it not be true that we experience the color of the ceiling, an off-white, say, or the taste and smell of a *macchiato* coffee, or the sound of a tree falling gently in the forest? These are all phenomenal properties. When I say that we "experience" phenomenal properties, I mean simply to generalize the verb that goes with the sensory modality in which the phenomenal properties appear. So, for example, we see colors, and we hear sounds, and so on, but I shall sometimes say that we experience colors, which admittedly sounds a bit odd, almost sensual, and that we experience sounds, and so on, meaning by "experience" nothing more than "see, or hear, or ...".

It would be more natural to call all the things we experience by their proper names ("colors," "tastes," "smells," "sounds"), and to say that we see and test and smell and hear them, but we can perhaps accept "phenomenal properties" as a generic name for all of these things, and no harm done, provided that we remember that we *are* talking about colors, or sounds, or tastes, or smells, or sounds or whatever is experienced, and that they are all very different from one another. One of the dangers of using the generic word "experience" is that we may think that there must be something common to seeing and hearing and so on, called "experience," perhaps in *addition* to seeing and hearing, and that we can catch ourselves in the act of experiencing *it*.

I find it difficult to think of anything that all the phenomenal properties, or whatever they are called, have in common, or what they all *are*. Colors are seen, certainly, and sounds are heard, smells are smelled, tastes are tasted, and so on, and, in the dangerous philosophical jargon that we have just accepted, they are all perceived or experienced. They are all the objects of experience, but of course that doesn't narrow things down much. Revolutions, skyscrapers, and lunar landings are all experienced, but so are bursts of anxiety, insincerity, and burglars.

Phenomenal properties, like qualia, are sometimes supposed to *be* experiences; but sometimes they are supposed to be *properties of* experiences.

Suppose that they are *properties of* experiences. What we experience are experiences, presumably, so if phenomenal properties are properties of experience, then we do not experience *them*. Suppose on the other hand that the phenomenal properties are the experiences themselves. If this is so, one is bound to wonder what the experience itself is like—the bare experience—without the property to ginger it up with content, and how it might differ from itself with and without a property. These are all silly options, no doubt, but it shows what a tangle one can get into with terminology like "phenomenal property," "qualia," "sense data," "sense contents," and other made-up words in philosophy, including "experience." Gilbert Ryle once gave a lecture in which he denied the existence of experiences.

During the discussion period afterward, Donald MacNabb asked, "How would it be, Gilbert, if I were to kick you on the shin?" "Yes, Don," replied Ryle, "that *would* be an experience."

There is something *grammatically* off about saying either that *a color* is an experience (not at all the same thing as saying that a color is experienced), or that a color is a property of an experience. On the first option, we are able to say that a color might last ten minutes, since it is an experience and an experience can last ten minutes, which is absurd. On the second option, we have to say that the color of my experience is off-white, or, worse, that my experience is off-white *colored*, rather than that I have an experience *of* an off-white color. We should no more say that an off-white color is a property of an experience, because we experience an off-white color, than we should say that an elephant is a property of an experience, because we experience an elephant. "Off-white" can be a grammatical subject, when it is a noun, just as much as "elephant" can.

My own view is that colors and sounds and the rest are neither experiences nor properties of experiences, nor phenomenal properties, if these are any different from colors and sounds, or qualia. It remains perfectly true that we see colors and hear sounds. Colors and sounds are the *objects* of experience, the "proper" objects of experience in fact, as philosophers have called them, not properties of experience at all. Suppose I see a nice red strawberry. Then the red color is the "proper" object of perception, and the strawberry itself is the physical object that we see *by* or *in* seeing the color. The strawberry is the object, but not in this parlance the "proper" object. One of the ways we can tell that what we are seeing is a strawberry is by its color, though three-dimensional shape is perhaps even more important. (On the other hand, what would we make of something shaped like a strawberry and *blue* in color?)

The tendency to locate colors and sounds and the other phenomenal properties in the mind, whether as experiences or as properties of experiences, has a long and troubled history, and I will give a brief account of a relevant part of it to orient readers to my own view, and to prepare them for what is coming in the final chapter on neutral monism.

The place to begin is with John Locke's well-known distinction, following the terminology due to his contemporary Robert Boyle, between *primary* and *secondary qualities*.[1] The primary qualities must not be confused with qualia. "Primary qualities alone" really exist in bodies in such a way as to "resemble" our perceptions or "ideas" of them, whereas the ideas of secondary qualities do *not* resemble the "powers" in bodies that cause them; or so Locke thinks. So the "ideas" of the Lockean qualities, *both primary and secondary*, are phenomenal properties, but the qualities themselves are not. For Locke, the character of the whiteness that we experience, the quale, resembles nothing in the white thing itself, unlike the character of the square that we see. The only connection is that the white thing will *cause* the experiences or qualia of whiteness, just as the square thing will cause the experiences of squarenesses. But in the latter case there is a resemblance, in the former not.

Locke gives a truly terrible argument for this conclusion about resemblances. The cause of whiteness does not resemble whiteness, because white disappears in the darkness, whereas the cause does not! Well, does it disappear in the dark? In the dark, it's hard to see, surely, so how would one *know* that it had disappeared? By a spectrophotometric check? Well, would it be a visual spectrophotometer, or one that records levels of radiation at particular wavelengths? If it is a visual spectrophotometer, one will see nothing, because the instrument doesn't work in the dark. The spectrophotometer itself will anyway disappear in the dark. If color *does* disappear, visually, in the dark, then so does everything else, visually, including squares. So the ideas of squares are the ideas of secondary qualities. Locke gives *lots* more bad arguments for the same conclusion.

However, it is important to know that Locke's distinction corresponds exactly to what was required by the emerging science of the time, the new science that was taking shape in the scientific revolution. The primary qualities are the mathematical ones, such as shape, number, and size. The secondary qualities, which are perhaps in a way more qualia-like, are colors, sounds, tastes, and so on. (Still, it is important to remember that when I see a square, remembering too that shape is a primary quality, then there is a square quale, or a square, as I should call it, just as there is whiteness if I see something white. For Locke, the difference between the two cases is that the idea of the primary quality *resembles* the quality, whereas the idea of the secondary quality does not.) The science of Locke's time looked to things like compression and rarefaction of groups of particles to explain sound, or the size of particles to explain hue and color, regarding the qualitative or experienced sound or color as the *idea* of a secondary quality, or, as Newton says, as the sensation of color. In the *Opticks*, published in 1704, Newton aligned large particles with red, smaller ones with blue. A barrier through which the particles can pass will deflect the large particles at a less acute angle than the small ones. Wave theories of light, such as Huygens's, also separated what we experience in the way of light from its physical essence.

During the scientific revolution the qualitative aspects of experience were being pushed into the mind in the scientific view of the way things are, leaving a mathematical world for science to investigate—very successfully, as it happened. The mind was conveniently understood as whatever was *left over* from the successful mathematical sciences. For example, geometrical optics with its laws of reflection and refraction and so on did not include colors.

During the scientific revolution ... the mind was conveniently understood as whatever was *left over* from the successful mathematical sciences.

Therefore they were in the mind and not in the physical world, except as tendencies to produce sensations of colors.

The conception of the phenomenal property and of the quale results from a world conceived, for the purposes of science, as separated into two very different and incompatible pieces: matter, consisting of linear dimensions or extension, and mind, consisting of consciousness, in Descartes's version, or sensations, in Newton's.

The word "qualia," however, was not introduced into philosophy until 1929, when it was first used by the American philosopher C. I. Lewis, but for him qualia were to be understood as properties of things called "sense data." These objects were notoriously difficult to define, but it is fair to say that they were experiences taken (if such a thing is possible) as *entities*. It is interesting that since Lewis's time the philosophical work that sense data did in mid-twentieth-century philosophy of mind and perception has been turned over to qualia. Like sense data, they have a special relationship to consciousness, in the sense that there can no more be an unsensed sense datum than there can be an unexperienced quale.

It is also interesting that when in 1910 G. E. Moore began to use the term "sense datum," which had been introduced in 1885 by Josiah Royce, Moore was unsure whether sense data were parts of the surfaces of material things. So his conception of sense data was not one in which they were necessarily either experiences or properties of experiences. They might have been entirely physical, and Moore found it very hard to decide on the answer to the question.

In spite of the difficulty of characterizing what one might call the *content* of our experience, or the things that we experience (colors, sounds, tastes, feelings, awareness of anger, uneasiness, pains, pangs of guilt, a feeling of imbalance, and so on) it is, I think, best to start from their undeniable presence, without trying to characterize them in a way that abstracts away from their ordinary and more accurate characterization as colors, sounds, and the rest. The abstraction—calling them all phenomenal properties or qualia, without being able to explain what this *means*— also severs them from their relationships with the other things we experience, and makes it hard to remember that they are the *objects* of experience; they are what we experience, not experience itself.

Let us therefore finally examine theories that do not use made-up words, and also accord equal weight to the physical world and to the contents of experience and mind and do not try to reduce one to the other. Most people do not find it hard to believe that there are mental things as well as physical things, even if they do not understand the *relation* between them—which is to say, even if they do not know the solution to the mind–body problem.

We should remember that dualism is really a neutral theory, in the sense that it favors neither mind nor matter but allows a place for both of them, though, as we have seen, it thereby creates the mind–body problem. What I want to look at next are theories that, like dualism, do allow a place for both mind and matter, including the body, but that do not create the problem, or, if they do, resolve or even dissolve it. The conclusion that we have reached so far is that things such as color, pains, the deliverances of the sense of balance, and all the rest of the things that philosophers today casually refer to as qualia or as consciousness, have to be included on the ground floor of any account of the mind–body problem. The reason is that they resist the reductive theories of mind, and over the last hundred years they have seen such theories off the field.

Dissolutionism

"Dissolutionism" about the mind–body problem is not really a single view, and nor perhaps would the word "dissolutionism" be widely recognized today, even among philosophers. The mind–body problem is to be *dissolved*, it is said, rather than to be *solved* by a philosophical or a scientific theory. The reference here is to the work of Wittgenstein, whose view was that *all* philosophical problems should be dissolved, rather than solved by large-scale philosophical theories. Wittgenstein himself did not devote a great deal of time to the mind–body problem as a formal problem, though the last part of his working life until his early death in 1951 was in large part devoted to thinking about mental and psychological concepts and their understanding. In a celebrated passage in the *Philosophical Investigations* he asks,

> How does the philosophical problem about mental processes and states and behaviourism arise?—The first step is the one that altogether escapes notice. We talk of processes and states and leave their nature undecided. Sometime perhaps we shall know more about them—we think. But that is just what commits us to a particular way of looking at the matter. For we have a definite concept of what it means to know a process better. (The decisive movement in the conjuring trick has been made, and it was the very one we thought quite innocent.)—And now the analogy which was to make us understand our thoughts falls to pieces. So we have to deny the yet uncomprehended process in the yet unexplored medium. And now it looks as if we had denied mental processes. And naturally we don't want to deny them.[2]

We might take a pain as an example. I feel a pain in my hand, say. This seems an unproblematic thing to say. When we try to conceptualize what has been said, we think of the pain as a *state* of the hand, or part of the hand, or some sort of *process* going on in the hand, perhaps having to do with damage to the hand. And here Wittgenstein is on very strong ground. We think we understand what kind of thing this state is, because we are comparing it to a physical state, say, a state of matter. This matter is liquid, that matter is solid; there is nothing hard to understand here. We think we have advanced our understanding when we say that the pain is a *mental* state, rather than a physical one. That state is physical, that one is mental. For Wittgenstein, however, the deeply problematic word here is "state," not "mental." How do we know that the pain is a state of the hand, and what does that mean? On the surface it hardly seems to say more than that I have a pain, and that it is in my hand. "Pain is a nonphysical state" seems to add little to that, and yet it *forces* the mind–body problem upon us. What is the relationship between the physical state, damage to the hand, and the mental state, pain? In view of the evanescent and impalpable character of the mental, one might wish to reduce it to some physical state, perhaps a behaviorist one, as Wittgenstein imagines in the passage above.

It seems, then, that Wittgenstein's difficulty has to do with the understanding of our first proposition, that the mind is a physical *thing*. Everything goes wrong when we think of the mind as a thing on the model of a physical thing, he thinks. And he is right. It is no better to think of the mind as a collection of physical *states*, such as pains or thoughts. What Wittgenstein's positive views are is for the students of his later writings to determine, but it is clear that "The mind is a nonphysical thing" and "The body is a physical thing" will not pass muster with him as a description of the two key elements that feature in the mind–body problem.

What is wrong with putting the two propositions alongside one another is the false parallelism that it sets up between mind and body. It is not so much that the proposition that the mind is a nonphysical thing is straightforwardly false; the difficulty is in working out what sense can be attached to it. This turns out to be no easy project. Here is the physical state, say, electrons buzzing around atoms, and other particles doing their dance, and here is the mental state, with a pain quale attached. This sort of "Heath Robinson" metaphysics immediately follows from something as innocuous looking as the first two propositions in the mind–body problem: the mind is nonphysical, and the body is physical. The difficulty, for Wittgenstein, is in reaching a commanding perspective from which to view the first proposition, and it was to this project that he devoted his later work on the philosophy of psychology.

The distorting parallel drawn between the nature of mind and of the body is also the theme of Gilbert Ryle's 1949 book on *The Concept of Mind*. Ryle agrees with Wittgenstein that there is nothing wrong with the ordinary use of mental concepts and words in the sentences we utter throughout our daily lives, but that it is another matter to understand what these words and sentences mean.

Ryle writes that we make a "category mistake" when we put concepts into "logical types to which they do not belong."[3] Things get interesting when the result is a mistaken conception of the objects of these concepts. Ryle's classic example concerns the Oxford colleges, which collectively make up the University of Oxford. He imagines a visitor to Oxford, who does not understand the fact that Oxford is a collegiate university, seeing all the colleges and then wondering where the university is to be found. The visitor might conclude, on not finding the university in any physical place in or near the city of Oxford, that it is instead to be found, an ethereal and invisible college, in a nonphysical place. The mistake such a visitor has made is to put the concept of the University into the same category as the concept of the colleges. Ryle calls this a "category mistake," and it occurs when something of one category is put into another category to which it does not belong. (Ryle's simile works perfectly well as it is intended, though it is not strictly accurate, since, for example, apart from the forty-four halls and colleges that comprise the University of Oxford, as such, it owns and administers various entities, such as the Botanic Gardens and museums that belong to no college, and it sets examinations and awards degrees, which no college does. So in some sense the University of Oxford *is* something over and above its colleges.)

Another interesting and more difficult instance of the category mistake can be found in the attempt to say, in the same "logical tone of voice," as Ryle puts it, that I have a left-hand glove and a right-hand glove, and that I have a pair of gloves.

> It is perfectly proper to say, in one logical tone of voice, that there exist minds, and to say, in another logical tone of voice, that there exist bodies. But these expressions do not indicate two different species of existence, for "existence" is not a generic word like "coloured" or "sexed." They indicate two different senses of "exist," somewhat as "rising" has two different senses in "the tide is rising," "hopes are rising" and "the average age of death is rising." A man would be making a poor joke who said that three things are rising, namely the tide, hopes and the average age of death.[4]

What is wrong is the coordinating conjunctions "and" in the example of the gloves? I do not have one thing and the other thing, two gloves, *and* a pair of gloves, but ... and here Ryle's powers of analysis weakened before the challenge. He found himself saying that just as having a right-hand glove and having a left-hand glove *is* having a pair of gloves, so that the "and" is out of place, so similarly having a body, and one disposed to do such-and-such, *is* having a mind. In this way Ryle fell into a kind of behaviorism that he himself later came to regret, and he spent the last part of his life up until 1976 trying to make up for it. He wrote a sequence of papers devoted to the topic of *thinking*, trying to show how on the one hand thinking is not a higher inner "state" at all, and on the other that it is not just a bodily activity. He was convinced, however, that it is also not something we do in addition to our everyday activities. He notes that even schoolchildren know what it means to be asked to *think about* something, and can do it if invited to. Yet in spite of its simplicity the logic of the concept evaded Ryle's best efforts, though his papers on the topic remain the subtlest and most clear-headed ones on the subject.

The idea of the category mistake, and its application to the mind–body problem, can certainly be pulled apart from behaviorism, even though Ryle never quite managed to do it. What we need to know is that, just as Ryle claimed, "His mind is nonphysical and his body is physical" embodies a category mistake. The sentence is a *zeugma*, the Greek for a "yoking together." Another amusing *zeugma* is "Give neither counsel nor salt till you are asked for it," as the proverb goes. The best-known example of a *zeugma* in English is "She came home in a flood of tears and a sedan chair." Council and salt are two radically different kinds of things, and it does not make complete sense to tie them together with a coordinating "and," and talk about them "in the same logical tone of voice." The same is true of tears and sedan chairs. They do not belong in one category (abstract things, or "things I came home in"?), and we have in the *zeugma* the mistake that Ryle railed against. The psychophysical claim does indeed almost sound like the joke that the *zeugma* is sometimes meant to be: "His mind and his body got out of bed together; both of them were ready for breakfast."

Yet Ryle picks on the wrong reason to criticize the mistake. What is wrong with saying that I have a left-hand glove and a right-hand glove and a pair of gloves, he thinks, is that the two gloves and the pair are *one and the same thing*. That may seem to be true in the case of the gloves, but it most certainly does not apply in the case of the mind and the body. "My body is stuck in the door" does not even *imply* that "My mind is stuck in the door," much less amount to the same thing.

The reason that "I have a mind and I have a body" is a *zeugma* is not that my mind and my body are one and the same thing and that the sentences involve some kind of silly repetition. "He has a good mind" and "He has a good body" have entirely different *senses*. The problem with "I have a left-hand glove and a right-hand glove, and I have a pair of gloves" (assuming that what is being said is not that I have some things like a red child's glove and a man's gray glove and a pair of lady's gloves) is that referring to the gloves individually puts them into one category (left- or right-handed things) and referring to them collectively puts them into another category (pairs of things that belong together). Then it seems that one can count what is in each of the two categories separately in one supracategory, and end up with the idea that I have *three* countable things in the supracategory: the two gloves *and* the pair. But there is or should be no supracategory, or else our arithmetic will go off the rails.

One can construct an inconsistent tetrad using Ryle's example.

(1*) The pair of gloves is nonphysical (because it is an abstract set).

(2*) The left-hand glove and the right-hand glove are physical.

(3*) The pair of gloves and the left-hand glove and the right-hand glove interact. If I tear the left-hand glove and I tear the right-hand glove—lesions!—then the pair of gloves is torn; and the reverse is true, as well.

(4*) Physical and nonphysical things cannot interact.

It is reasonable to think that the big mistake is the conjunction or "yoking together" of (1*) and (2*). The conjunction means that there are two things, the pair *and* the left-hand glove and the right-hand glove, and they belong to two kinds of existence, one physical and one nonphysical.

The difficult thing to understand is that pairs do exist, and that a pair of gloves is not at all the same thing as a left-hand glove and a right-hand glove, even if the two gloves *are* a pair. It is of course true that the pair could not *physically* exist without the two things that make it a pair, but that does not mean that the pair and the things are one and the same anything. To what common kind, we can inquire, do gloves and pairs or abstract sets belong? Why is it so wrong to "yoke together" mind and body in the way that our first two propositions do? Suppose that someone says, "His mind and his body are in the house." This is certainly wrong. "His mind is in the house" commits a category mistake, even if he is in the house, though not the mistake that Ryle imagined. It treats one thing (the mind) as if it belonged to a category to which it does not belong (things in space). But it is just as misleading, or even more misleading, to say, "His body is in the house, but his mind is not," all in one breath, because that seems to imply that his mind really *might* have been in the house, but just happens not to be at the moment. The proposition that his mind is in the house in not merely false; it is rather that there are no circumstances under which it *could* be true, since there is nothing that the words *could* mean or come to mean, and no reason that could present itself to us that would force us to decide that it is true.

Sets and other mathematical objects exist, but there are no circumstances in which it *could* be right to include gloves among them. Mathematics is about mathematical objects and structures, such as pairs, but not about physical things, such as gloves. In the mind–body problem, the category mistake is the mistake of trying to put nonphysical, nonspatial things alongside physical spatial things in a causal sequence, and physical spatial things alongside nonphysical, nonspatial things in a causal sequence.

The Double Aspect Theory and Panpsychism

There are two more theories that commit the kind of category mistake that I have been describing, in spades, and then, stubbornly refusing to give it up, build it into the structure of the world. This is a dispiriting thing to see in philosophy, and the kind of thing that gives metaphysics a bad name. It is certainly a good thing that both the double aspect theory and panpsychism refuse, with equal obstinacy, to deny that colors and sounds and tastes and the other things collectively called qualia do exist, and that they can be reduced to things other than themselves. That does not give the proponents of these theories the right to continue to commit the category mistake.

The double aspect theory has Spinoza in the seventeenth century as its main standard bearer, and there are elements of it in the work of Thomas Nagel in our own time. According to Spinoza, there is only one substance or ultimately real thing in the universe, but it can be viewed under two complementary aspects: extension or linear dimensions, and thought. The same thing is true of the human being. It can be viewed in two ways. It can be viewed under the aspect of thought, as a mind; or it can be viewed under the aspect of extension, as a body. There are not two things here, however, and accordingly they cannot interact. Mind and body no more interact than a book classified by price, say as a $25 item, interacts with *the same book* classified by subject, say, as a work on astronomy. There are not two books that interact, an economic book and an astronomical book, but one book, classified linguistically in two ways: by price and by subject. The double aspect theory works by denying that mind and body interact. It remains a puzzling thing, however, why the ultimate constituents of reality should have a double aspect and manifest themselves as both thought and extension, two things that Descartes regarded as *incompatible*.

True, thought cannot be reduced to extension, just as redness cannot be reduced to long-wave light. We can say that pink is a light red, but we cannot say that pink is a light long-wave light. That does not mean that we should say that everything having the color red can be regarded either as red, under its phenomenal aspect, or as producing

Mind and body no more interact than a book classified by price, say as a $25 item, interacts with *the same book* classified by subject, say as a work on astronomy.

long-wave light, under its physical aspect. Or if we do say that, we should be ready with an explanation of the relationship between the two aspects. Why should red ally itself with long-wave light particularly? What is the connection? Without some explanation, the double aspect theory simply reproduces the mind–body problem in *all* parts of existence. The same is true of panpsychism, the view that mind, or even consciousness, is a property of everything in the universe. The two theories give no proper account of the way in which mind and body are causally related. It is no answer to the question how the king and queen are related to say that they are both inevitable and double *aspects* of the monarchy, or that, appearances to the contrary, it is not surprising that they are king and queen because, appearances to the contrary, every part of existence has its kingly and queenly aspect. This is unconvincing, and at least we need to know more about the relation between the two aspects.

There is a view called "panprotopsychism" that is designed to help with this final problem, but it does so not at all. It claims that throughout nature and existence is an unknown something, an *x* factor, and *x* everywhere exhibits on the one hand mind-like or nonphysical aspects and on the other physical aspects, including bodily ones. This is what one might call a bodge job, whose only virtue is that it solves the problem created by panpsychism.

Panprotopsychism agrees that panpsychism leaves us with the mystery of how the physical and the nonphysical are related. It gives the answer that both of them are derived from something more fundamental, x. But it does not tell us what x is, only that it is, or *might* be, and that it *might* produce the incompatible properties of space or the physical and consciousness.

NEUTRAL MONISM

Introduction

The double aspect theory discussed in chapter 6 is often confused with a much more powerful and very different theory known as neutral monism. The last well-known proponent of neutral monism was A. J. Ayer, writing in 1936. Earlier the view had been advanced by Bertrand Russell (between 1919 and 1927) and William James and the American new realists, but first and most powerfully by Ernst Mach. Mach's fullest exposition is contained in *The Analysis of Sensations*, published in German in 1885 and then in English in 1897.

The view of the neutral monist is that neither mind nor matter is basic, and that both are composed of more basic neutral elements, elements that are in some ways very similar to qualia. Qualia-like things, or phenomenal properties, make up the world, rather than the world, or

The view of the neutral monist is that neither mind nor matter is basic, and that both are composed of more basic neutral elements, elements that are in some ways very similar to qualia.

the brains of human beings, producing qualia. "Bodies do not produce sensations, but complexes of sensations (complexes of elements) make up bodies."[1]

Recently there has been a revival of interest in this kind of view, as there has been in panpsychism and pan-protopsychism. The revival of neutral views, including neutral monism, is in part due to the recalcitrant nonphysicality of phenomenal properties. They have resisted the best attempts of the twentieth century to analyze them away.[2] Proponents of the neutral views, though, have recently been tending to interpret the phenomenal elements of neutral monism, such as colors and sounds, as concrete dynamical events in the world described by physics. Eric Banks, for example, writes that "qualities are simply the concrete manifestations of powers in events, observed or not, that occur around us all the time."[3] This is a view that differs hardly at all from Locke's representational realism, the view that what we are aware of is not the real external world, but a representation, a mental image or sensation or perception. And this mental image, sensation, or perception is supposed by Banks to be a dynamic *physical* event in the brain.[4]

In fact, neutral monism takes as its basic elements genuinely neutral things such as "colours, sounds, temperatures, pressures, spaces, times, and so forth,"[5] and takes both the self and the physical object as only more or less permanent collections of these elements. It is true that Mach sometimes rather misleadingly also calls his elements "sensations," but he states quite clearly that he means not that they are psychological in nature, but that their nature is what turns up *in* sensation. They are psychological in nature when they are placed in a psychological causal sequence; otherwise, they are what we would ordinarily regard as the perceptual or qualitative *content* of sensations.

We cannot accept Russell's own claim that he continued to be a neutral monist after 1927, and nor can we accept his claim that during this period he was not the kind of representative realist who took sensations to be part of the brain. In reality Russell gave up neutral monism in 1927, in favor of a view in which qualia are properties of events in the brain. But the brain is *itself* physical, or rather, the neutral brain *elements*—the images and qualia of the brain in all the perceptual modalities—are placed in the physical sequence and are, therefore, physical. Consequently, they cannot in any sense *contain* the neutral elements that are placed in a psychological sequence.

Banks writes, on Russell's behalf and for himself, that "I now see that enhanced physicalist view of the world involving events and natural qualities comes first, before the mental in every sense. ... Sensation qualities are just the higher-order qualities of very complexly configured events in our brains."[6] How is this *neutral* monism? It is a *physicalist* monism, "Russellian monism" as it is sometimes called now, though it is doubtful whether Russell himself ever held it, a view that *ignores* the qualitative in favor of the abstract and the physical. Russell's view about the mind–body problem after 1927 was a form of central-state materialism, combined with representative realism.[7] And in Banks's description we also have a strong suggestion of a kind of emergentism. The qualia are "higher-order," and the events are "very complexly configured," which is somehow supposed to help in allowing the qualia to inhabit them.

The upshot of Mach's view is that we should not conceive of perception as a transaction between things in the external world—things of which we are not directly aware—and a self, which then blossoms with sensations, of which we *are* directly aware: representative realism.

> For us, therefore, the world does not consist of
> mysterious entities, which by their interaction with
> another equally mysterious entity, the ego, produce
> sensations, which alone are accessible. For us,
> colors, sounds, spaces, times … are provisionally the
> ultimate elements, whose given connection it is our
> business to investigate.[8]

Here as elsewhere I see only a phenomenalistic Mach, and my view is that the logical positivists who also interpreted Mach in this way were correct.[9] It cannot seriously be doubted that Mach wanted the contents of unmediated experience (the things we see directly, hear directly, and so on, such as colors and sounds) to be his elements. It is easy to see how instances of colors and sounds can be taken neutrally, either physically or mentally. The same is not true of dynamical forces or Lockean powers to produce colored sensations in us, which are then to be identified with brain processes, especially since these brain processes are not in any literal sense colored or auditory or possessed of any of the other proper objects of the various sensory modalities.

For Mach and the Russell of 1919 to 1927, perception is, like everything else we know, a relation between the elements. Mach's elements include, very significantly, "spaces," and it follows that in themselves none of the elements is spatial. Space and position elements are associated according to Mach with the motor movement of the eye. Mach devotes two chapters, amounting to twenty percent of *The Analysis of Sensations*, to sensations of space, and to other closely related topics such as relative position and changes of shape with orientation.

> I once heard the question seriously discussed, "How the perception of a large tree could find room in the little head of a man?" Now, although this "problem" is no problem, yet it renders us vividly sensible of the absurdity that can be committed by thinking sensations spatially into the brain. When I speak of the sensations of another person, those sensations

are, of course, not exhibited in my optical or physical space; they are mentally added, and I conceive them causally, not spatially, attached to the brain observed, or rather functionally presented. When I speak of my own sensations, these sensations do not exist spatially in my head, but rather my "head" shares with them the same spatial field.[10]

There are, of course, sensations that *are* spatially as well as causally associated with our bodies, such as pain and hunger, but color and the other elements are experienced outside the body, if experience is to be our guide. I do not *see* the green of the tree's leaves inside my hand or my eye. All the elements of the external world (the world external to our bodies), the elements of our bodies, and those we take to be nonphysical form a single mass, internally connected in various ways. "In this way, accordingly, we do not find the gap between bodies and sensations described above, between what is without and what is within, between the material world and the spiritual world";[11] for there is no gap.

Neutral Monism: The Theory

The most important tenet of neutral monism—what makes it genuinely neutral—is that a neutral element, considered in a physical sequence, *is* physical, but the very same element, considered in a mental sequence, is *on that account* to be regarded as mental:

> Thus the great gulf between physical and psychological research persists only when we acquiesce in our habitual stereotyped conceptions. A color is a physical object as soon as we consider its dependence, for instance, upon its luminous source, upon other colors, upon temperatures, upon spaces, and so forth. When we consider, however, its dependence on the retina [and other elements of the body], it is a psychological object, a sensation. Not the subject matter, but the direction of our investigation, is different in the two domains.[12]

Consider again Ryle's simile of the University of Oxford and the colleges that make it up. Because the colleges and the university are different *kinds* of things, we cannot put them into one category and then proceed to count them. We cannot ask how many things there are in the following list of institutions: Exeter College, Trinity College, Balliol College, University College ... (and all the other halls and colleges), *and* the University of Oxford. The list itself involves a category mistake. In just the same way, there is a mistaken categorization in the following list of *gloves*: a left-hand glove, a right-hand glove, a red glove, another

The most important tenet of neutral monism ... is that a neutral element, considered in a physical sequence, *is* physical, but the very same element, considered in a mental sequence, is *on that account* to be regarded as mental.

red glove, a child's glove, another child's glove ..., *and* three pairs of gloves. For neither a pair of gloves nor a number of pairs of gloves is a glove.

In just the same way, we can consider a sequence consisting of neutral illumination, a red surface, changes in electromagnetic potential, and so on. If we think of this very same red element in a sequence that includes blinking, human expectations, the color surround of the red element, the state of the retina, lack of damage to the area V2 in the visual cortex, and so on, then the red element is to be regarded as a psychological or perceptual event.

What we cannot or should not do is create a sequence of physically interpreted elements containing a psychologically interpreted element, or a sequence of psychologically interpreted elements containing a physical element. The construction of the set of events that includes blinking, human expectation, the color surround, the state of the retina, lack of damage to V2, a perception of red, *and the red physical surface* makes a monumental category mistake, just as the construction of the set consisting of a neutral illumination, a red surface, changes in electromagnetic potential, *and a perception of red* makes a mistake of principle. Neither can be admitted as a legitimate causal sequence.

One of the more dramatic consequences of Mach's monism is that such things as pains, taken in the physical sequence, are physical and not mental, as Descartes took them to be. I imagine that this view will be thoroughly unacceptable to almost everyone except for those of the medieval and earlier philosophers for whom sensations were *physical*, and of course Mach himself. It has the consequence that the distinction between the physical and the mental or psychological is not to be drawn where the study of physics and psychology placed it in Mach's time, or where it is placed today. That is what Mach intended. Pains are physical things in the body that have spatial locations and can be associated with visual spatial locations, but they are not part of the physics of electricity and magnetism, of optics and acoustics and so on. Why should this disturb us?

Bertrand Russell describes neutral monism with the memorable image of an old-fashioned postal directory:

> "Neutral monism"—as opposed to idealistic monism and materialistic monism—is the theory that the things commonly regarded as mental and the things commonly regarded as physical do not differ in respect of any intrinsic property possessed by the one and not by the other, but differ only in respect of arrangement and context. The theory may be illustrated by comparison with a postal directory, in which the same names appear twice over, once in alphabetical and once in geographical order; and we may compare the alphabetical order to the mental

and the geographical order to the physical. The affinities of a given thing are quite different in the two orders, and its causes and effects obey different laws. Two objects may be connected in the mental world by the association of ideas, and in the physical world by the law of gravitation. The whole context of an object is so different in the mental order from what it is in the physical order that the object itself is thought to be duplicated, and in the mental order it is called an "idea," namely the idea of the same object in the physical order. But this duplication is a mistake: "ideas" of chairs and tables are identical with chairs and tables, but are considered in their mental context, not in the context of physics.[13]

From the neutral monism of Mach and Russell we can take the idea of the two intersecting sequences of objects of inquiry, physical and psychological. From Ryle we can take the idea of the category mistake applied to these sequences. The mistake that produces the mind–body problem comes, just as Ayer and Ryle claimed, from concocting sequences of perceptual or psychological events ending in physical events, or vice versa, sequences that embody the category mistake at their inception. From Mach and the other neutral monists we can take the background understanding of what the two distinct sequences, mental and physical, must be, and why the category mistake is indeed a mistake, in the context of mind and body.

The real sticking point in the mind–body problem is to think simultaneously that the mind—all of it—is nonphysical and that the body—all of it—is physical. That would be like saying (i) that the left-hand glove and the right-hand glove are physical, and (ii) that the pair of gloves is nonphysical, because it is abstract. Either of these propositions can be asserted singly, but not at the same time as the other. To assert the two propositions simultaneously produces the paradox that when I have a pair of gloves I have *three* things, two physical and one nonphysical, which is absurd. Similarly, to assert that the mind is nonphysical and the body is physical produces a paradox: the mind–body problem. We should keep the two sets of accounts distinct, realizing that this does not prevent the mind and the body from interacting. As Moritz Schlick writes,

> The so-called "psycho-physical problem" arises
> from the mixed employment of both modes of
> representation in one and the same sentence. Words
> are put side by side which, when correctly used, really
> belong to different languages. This gives rise to no
> difficulties in ordinary life, because there language
> isn't pushed to the critical point. This occurs first
> in philosophical reflection on the propositions
> of science. Here the physicist must needs assure

us that, for example, the sentence, "The leaf is green" merely means that a certain spatial object reflects rays of a certain frequency only: while the psychologist must needs insist that the sentence says something about the quality of a perceptual content. The different "mind–body theories" are only outgrowths of subsequent puzzled attempts to make these interpretations accord with one another. Such theories speak for the most part of a duality of percept and object, inner-world, outer-world, etc., where it is actually only a matter of two linguistic groupings of the events of the world. The circumstance that the physical language as a matter of experience seems to suffice for a complete description of the world has, as history teaches, not made easy the understanding of the true situation, but has favoured the growth of a materialistic metaphysics, which is as much a hindrance to the clarification of the problem as any other metaphysics.[14]

What should be said about the third proposition, which is that mind and body interact, from the point of view of Machian neutral monism? Here too "Mind and body interact" commits the same category mistake that we must be careful to avoid in conjoining the first two propositions. "The mind interacts with the body, and the body interacts with the mind" and "Mind and body interact" are sentences that embody the category mistake. "Words are put side by side which, when correctly used, really belong to different "languages." This gives rise to no difficulties in ordinary life, because there language isn't pushed to the critical point." The recognition of the category mistake forces us to take language "to the critical point" and beyond, and it seems that we cannot say things such as "The wine made my mind feel tipsy," without committing the category mistake.

Interaction, with Perception as the Example

The details of the interactions between the mental and the physical are tricky to manage in neutral monism, yet they can be managed, and that is part of the real wonder of the mind–body problem. Apart from Mach, the neutral monists have not given any consideration at all to the details of how such interactions work. I want to consider three examples of the interaction, of three different types, to illustrate the way in which the neutral monist should understand the relationship of causation between mental and physical events.

To begin, let us imagine that we see some external object in the ordinary way. Let us imagine that we see our own hand in front of us. We have the hand, and we have the image of the hand; however, the concept of an image of something is yet to be understood. The hand is a member of a physical sequence; the image is a member of a psychological sequence. There is an intersection between the two sequences, and what can be placed in the psychological sequence can also be placed in the physical sequence. The neutral monist can assert that *when the two sequences intersect, in the sense that there exists one element that can be placed in either sequence, we have the relation of causality between the physical and the psychological*. We have causality beyond "the critical point" of language referred to by Schlick. When we *do* descend to the level of the individual elements, however, we can understand mind–body interaction. At the level of the elements, the two kinds of sequence intersect, in the sense that there are elements that can be placed in both sequences; and there we have interactions. For this to occur, however, the physical elements must be capable of being taken as nonspatial, or the psychological elements must be capable of being taken as spatial. At this level it is the conjunction of the *first two propositions that is false* in our original inconsistent tetrad. One of them is false. Dualism is false, and monism is true.

From the point of view of what we are aware of, we have the sequence: <light>, <hand>, <hand reflecting light>, <light striking retina>, <activation of the visual cortex>. These elements can be classified as instances of the processes and objects of optics, anatomy, opto-electronics, physiology, and physiology, in that order. That is a physical sequence. We also have the sequence of images of <table>, <hand>, <grandmother>, <spaghetti>, <hunger>. These elements can be classified as instances of perception, perception, memory, memory, and desire. The sequence is psychological. But the two sequences intersect.

```
          <light>
<table>,  <hand>,  <grandmother>,  <spaghetti>,  <hunger>  (images)
          <hand reflecting
          light>
          <light striking
          retina>
          <activation of the
          visual cortex>
          (physical processes
          and objects)
```

The mind is to be taken as the sum of its parts, as Mach insisted; elements of desire, vision, memory, anticipation, and so on, and the relevant parts are capable of being "neutralized," inserted into a physical sequence, and taken as physical. The body, too, or parts of it, can be stripped down to the neutral elements, which can then be inserted into a psychological sequence. What we are left with, then, is the fact that the third proposition (that mind and body interact) is true only if we are prepared to take a mental or physical event, position it as a neutral event, and thence as a physical event, and vice versa; and the fourth proposition is true *simpliciter*.

There are also other ways of creating sequences of the same type. We could also regard the element <hand> as psychological, because it is placed in the *vertical* sequence of elements formed of events relating to the physical body, such as <activation of the visual cortex>.

As to "the mind is to be taken as the sum of its parts," there is nothing inherently difficult about this "reductionist" view. The neutral monist sees no advantage in thinking of mental states coming together in one mental place, the so-called Cartesian theater. David Chalmers has described the approach this way: the neutral monist and those who split the self up into its various parts "deflate the subject, either by denying that experiences must have subjects at all, or at least denying that subjects are metaphysically and conceptually simple entities."[15] One may not have to go all the way to the first "no-self" view, because the sum of the mind's parts can be regarded as the mind, but one does have to recognize that when the mind acts, or the body, it is not the whole mind or the whole body acting, as it were, concentrated at a point, but only one part of it. When I go to have lunch because of a feeling of hunger, it is the feeling of hunger, not the whole of my mental life, whether "deflated" or not, that takes me off to the diner.

Each of the first two propositions may be false, in the following way. Parts of the mind can be taken to be non-physical, and parts can be taken to be physical, and the same is true of the body. When mind and body interact, one of two things happens. Either the relevant parts of the mind or events in the mind can be given spatial characteristics, and can then interact with the spatial body; or the relevant parts of the body and the events in the body can be stripped of their spatial characteristics and can then interact with the nonspatial mind.

Interaction, with Sensations as the Example

The relationship of the properly subjective sensations (e.g., pains, aches, seasickness, a feeling of a scratchy skin) to the body may seem harder to understand than the relationship of identity in the case of the part of the physical object by which we see it and the image, which can be thought of in a literal way as having a place within the totality of consciousness.

A stomachache, to take an example, seems to have little in common with the physical stomach. How then can there be an overlap between the two sequences, such as to allow us to say that what causes the ache is the stomach? For one thing, the stomach is to be found in physical visual space, but apparently the stomachache is not to be found in physical space. There appears to be no member common to the physical and psychological sequences; there is no member that can be interpreted both in a physical and in a psychological way.

This appearance, however, is delusive. The scientific orthodoxy today is that the stomachache is not in the stomach, but in the head, in the firing of some neurons perhaps. This view is a very peculiar one, even if the neurons are very well integrated informationally or whatever. The reality is that there is a pain space, an ordering of the pains in a definitely vague and dim spatial organization, which maps onto "physical" space, a combination of the visual and tactile spaces. In this mapping the earache is located in the ear, the toothache in the tooth, and the stomachache in the stomach. Some pains and aches are harder to place, and they seem to move around. Some pains have vague locations. Early manifestations of appendix pain can masquerade as stomach pains, for example. But the location of the pain is not so vague that it can manifest itself in the ear, for example, or a finger, and still count as the pain that it is.

However, if we are careful, we can establish a distinct and different phenomenology for each of the kinds of what is commonly called "stomachache." We are not confronted with two blank and barely defined things: the otherwise undescribed mental pain, on the one hand, and the physical stomach, all of it, just sitting there, like a lump, on the other. The stomach is active, and there are numerous different kinds and causes of abdominal pain.

Among the causes of abdominal pain we have: ulcers; gallstones; pain from the appendix; menstrual cramps; indigestion; Crohn's disease; infection of the urinary tract; and many others. The felt symptoms of all of these are different from one another. But so are the detailed internal signs, if we look for them, in the abdomen and in general physiology.

The pain from ulcers is a gnawing, searing, and burning kind of pain, with some resemblance to hunger. It is to be found quite high in the front of the body, reaching from the bottom of the stomach up to the breastbone. It can also be mixed with a feeling of nausea, and perhaps bloating, especially after meals. The position of the feeling of bloating will also correspond exactly to the dimensions of the distension of the abdomen.

Gall bladder pain, on the other hand, may extend toward the right shoulder, and can feel dull and cramp-like, though sometimes sharp, and increasing with breathing in. And so on. The phenomenology of the different pains is very different, though all, of course, are pains. If we attend to it carefully, however, it becomes increasingly obvious that the different kinds of abdominal pain overlap in a precise way with their physical causes.

With ulcers, we have only to look at the sores in the stomach lining to understand more about the kind of pain that is suffered, and inspection of the sores might lead one to an understanding of the imbalance of stomach acids that can be a cause of ulcers. It is the sores or perforations that cause the pain, and the pain is in the sores. They certainly *look* sore, which is why they are called "sores." This is more than a learned association. It is a piece of phenomenology.

The element with ulcers at which a physical and psychological sequence cross is a searing located between the stomach and the breastbone. "Searing" means both the visible and felt fiery aspect of the sores, and the scorching that is caused by the stomach acids. For Descartes it was essential to override the ordinary language with which we describe the psychological pain and the physically painful condition, by means of arguments to the effect that we can conceive the psychological element without the physical element. But this kind of argument is in the end unrealistic. It is perhaps logically possible to have the pain of ulcers without ulcers, but what does this tell us? If such a strange condition were to occur, things would not be as they seem, and we would have no right to apply the ordinary psychological and physical criteria for identifying the pains. Are they really scorching pains? Are they properly localized? Is there the gnawing feeling present? And if the answers are still all affirmative, we should I think look for a physically interesting stomach condition that *mimics* ulcers, rather than concluding that the mind and the body do not interact. Nor should we conclude that they interact dualistically, in the sense that there is no logical, structural, or phenomenological overlap between the physical and the psychological.

Some recent research is going in this direction in the most interesting way. Martyn Goulding and his team at the Salk Institute have described the surprising discovery that the spinal neurons implicated in the tingling of a light touch are not the same neurons as those that relate to the pain ("chemical") itch, such as one due to a mosquito bite. The hope is to provide insight into the treatment of the chronic itch, because there is a neural pathway devoted to the pain itch.[16] Astoundingly, itching has its own complex physiology that is not the same as the physiology of the light touch. It is to be hoped that science will advance to a more and more specific understanding of the physiology of sensation in this sort of way, and I believe that the physiology and the psychology will move closer and closer and eventually converge, as they already do in the case of ulcers and many other examples in the other sensory modalities.

One might wonder how it can be that two such supposedly different things as sores and sorenesses can interact, or even be the same thing, or how the same thing can be represented by two senses, or by one sense and by thought. To answer this, we should consider what Leibniz says about Molyneux's problem. The problem is whether a man born blind, who has handled a cube and a sphere, would on regaining his sight, be able to tell by sight alone which was the cube and which was the sphere. Leibniz's answer is that the newly sighted man *would* be able to tell which was which:

> I am not talking about what he might actually do on the spot, when he is dazzled and confused by the strangeness—or, one should add, unaccustomed to making inferences. My view rests on the fact that in the case of the sphere there are no distinguished points on the surface of the sphere taken in itself, since everything there is uniform and without angles, whereas in the case of the cube there are eight points which are distinguished from all the others. ... These two geometries, the blind man's and the paralytic's [to whom touch is denied] must come together, and agree, and indeed ultimately rest on the same ideas, even though they have no images in common.[17]

It is the same with the geometry of pains and the geometry of the physical body. They "rest on the same ideas [concepts], even though they may have no images in common."

Interaction between Thought and Action

Finally, even in the most extreme examples of pure thought in the mind, we can find an overlap between the psychological sequence of thoughts, and the physical sequence that includes the resulting action. Suppose I raise my right arm high, doing so because I believe that I have the right answer, and because I want the teacher to call on me. As a schoolboy, my arm has been trained over the years to rise, to shoot up, in the right way, palm forward. The thought that I have the right answer enters my conative and action-oriented consciousness, and here I am phenomenologically aware of it flowing into my "arm consciousness." The hardness and the strong pressure on my elbow on my desk seem to vanish. I have fingertip feelings above my hand, a sort of dancing toward the ceiling, and also a feeling that I know the right answer, and the answer is located *in the palm of my hand*! This is certainly very strange, and other people may have completely different but equally strange sensations, or none at all. I also have a strong awareness of my teacher's face, especially his eyes. Up goes my arm, with my fingers wiggling toward the ceiling. Will I be called on to give the wonderful right answer, the answer that is *mine*?

Now if we filter from all this phenomenology what is physically relevant, which is to say what can be located in space, we find the Machian element of *the arm shooting up, with sudden considerable acceleration, a snap in the elbow, and the fingers twitching at the ceiling*. What do we find on physical observation of my arm? *The arm shooting up, with considerable acceleration, a snap in the elbow, and the fingers twitching at the ceiling.* Our two Machian elements coincide. Of course, we have to screen out much in the physical elements to see the match, and the same goes for the psychological elements; but the match is there. And when it is made, we know that we have the causation of the physical action by the psychological event: mind–body interaction. Such interaction can occur, because mental events can "become" or rather be taken to *be* physical elements, via their corresponding neutral elements. Which they are taken to be is a matter of which causal sequence they enter into. The first proposition in the inconsistent tetrad forming the mind–body problem is false, in the case under consideration. Here the mental is physical.

None of this means, however, that there are not elements "which obey *only* physical laws (unperceived material things, for example), some which obey only psychological laws (namely images, at least), or 'wild particulars,' as Russell called them, and some which obey both (namely sensations). Thus sensations will be both physical and mental, while images will be purely mental."[18] There could certainly be disagreement about which the "purely mental" elements are, but all we need in order to solve the mind–body problem is that the interacting elements, whichever they are, can be assigned to the physical or to the mental sequence, not that *all* elements have this character. There are images, in Russell's example, that simply refuse a placement in a spatial scheme, and there are others that do not.

Mental events can "become" or rather be taken to *be* physical elements, via their corresponding neutral elements.

There are also other images that can be placed at will in both sequences—for example, "floaters," the distorting wormlike images caused by condensations of the vitreous humor in the eye.

A Model for the Mind–Body Problem

To close, I want to develop a model of the inconsistent tetrad with which we started. This model will allow us to formulate the solution that neutral monism gives to the mind–body problem.[19] Imagine six refrigerator magnets, each with the shape of a numeral from "1" to "6," with all six numerals represented, and each having one of six colors, red (R), orange (O), yellow (Y), green (G), blue (B), and violet (V), with all of the familiar six colors represented.[20] Thus the elements of the array in the "refrigerator world" are colored numerals. The front of the refrigerator looks like this:

1R

3Y

5B

4G

6V

2O

Let these elements be the only things the refrigerator world contains. It contains only colored numerals. The colors and the numerals themselves are secondary. The primary things are the elements: colored numerals, not colors and numbers in the abstract.

However, we can arrange our elements in two very different sorts of sequence: mathematical and nonmathematical. The most familiar mathematical one goes 1, 2, 3, 4, 5, 6, and the most familiar nonmathematical one, the spectrum, goes R, O, Y, G, B, V. Now we can ask about our model world, in the quaint idiom of the scientists, "How does the nonmathematical element in a nonmathematical sequence 'arise' from the mathematical one?" How can we get colors out of numerals? Well, the short answer is that we don't, because we can't.

Suppose we find a sequence 1, 2, 3, 4, 5, V, for example. This is obviously impossible, as 1, 2, 3, 4, 5, V is not any sort of sequence, and we have the "numeral-color problem." But it is also a category mistake to place a color in a sequence of numerals. If you want the next term of a sequence represented by 1, 2, 3, 4, 5 ..., it may not be 6, but it cannot be V.

We can now create an inconsistent tetrad in our refrigerator world, numbered (1r) (the "r" for "refrigerator") to (4r), with the first and second propositions expressed as plurals. Mach was right to see the importance of *not* regarding the mind as a substance, uniformly physical or nonphysical. When the mind or its analogue in the r-tetrad is split up into its Machian elements, we can consider them separately, as in (1r), (2r), and (3r), below. It is important to develop this tetrad, as the neutral monists for the most part have contented themselves with vaguely thinking that the mind–body problem would go away merely if we were to think of all the elements as neutral. (Russell, writing in "On Propositions," is an honorable exception.) Here is the new tetrad.

(1r)   The color violet [an analogue for a part of the mind] is a nonmathematical [nonphysical] thing.
(2r)   The numeral 6 [an analogue for a part of the body] is a mathematical [physical] thing.
(3r)   Violet [a part of the mind] and the number 6 [a part of the body] follow [an analogue for causation] one another.
(4r)   Mathematical [physical] and nonmathematical [nonphysical] things cannot follow one another.[21]

It is important that in this analogy, the tetrad (1r)–(4r) refers to individual elements, in the plural (the colors, such as 6, and the numbers, such as violet), and not to two things representing "the mind" and "the body." (These analogues would be "color" and "number," both, as we say, "in the abstract.") For Mach and most neutral monists the right approach to the mind is the deflationary one. It is reductionist, in the sense that it reduces the mind to its elements.

If we substitute "the violet number (or numeral)" for "the color violet," it becomes immediately apparent that we *can* assign the neutral element (the violet number or numeral 6) to the mathematical sequence (1R, 2O, 3Y, 4G, 5B, 6V), in which case (1r) is false, or we can assign the colored numeral 6 to the nonmathematical sequence (R1, O2, Y3, G4, B5, V6), because it is violet, and (2r) is false. In both cases, the "color–numeral problem" has been solved. If we apply the lesson learned to the original tetrad about the mind–body problem, then the mind–body problem has been solved as well. The desire for coffee is usually placed in the psychological series. If it is, it makes no sense to also place at it the end of a *physical* causal chain, any more than it makes sense to put V at the end of the sequence 1, 2, 3, 4, 5 ... . But if we regard the desire as a physical one, and place it in the appropriate place in a physical sequence, we have mind–body interaction, for the desire is now to be regarded as a physical one, given a rough location in the body, and then given its appropriate causes and effects, including my reaching out for the cup of coffee.

We can even represent the different standard philosophical positions about the mind–body problem within the refrigerator world as attempts to make the tetrad consistent by rejecting one of its constituent propositions. For example, behaviorism states that a color is to be analyzed in terms of the disposition of a numeral to fall in a specified group of numerals; central-state materialism states that a color is identical with a numeral; functionalism states that a color is a functional or computational state of a machine, a machine that computes colors from numerals; eliminative materialism states that a perfected science will eliminate color descriptions in favor of the mathematically superior numerical descriptions; dualism states that colors and numerals are distinct entities. The last position is of course true in the real world, but false in the refrigerator world. With some ingenuity, the main existing scientific accounts of consciousness can be represented in a similar manner.

We do well to remember that the mind–body problem really is a paradox. Its solution is to be found in the intricate arguments for the four propositions in the inconsistent tetrad, and in the concepts embedded in these propositions. Neutral monism allows us to see this point very clearly indeed. We are used to thinking of the tetrad and the mind–body problem in fixed concepts, of the mental and the physical, and of the mind and the body. Then there is nothing for it; something has to give. Yet it is no good trying to wriggle out by assimilating one set of concepts to another, so that everything mental is declared to be really, incomprehensibly, physical, for example. The beauty of neutral monism is that it allows us to shift our given elements from one category into another, in a way that is legitimated by the phenomenology and does nothing to undermine the integrity of the given categories. Everybody knows that somehow the mind–body problem calls for a very fundamental shift in our understanding, and this is it. Our concepts must change gear, and neutral monism shows us how to do it. The pain of an ulcer is mental, in the sense that it can be scaled in entirely in psychological terms, and without a spatial reference, by its intensity, duration, quality, and felt location; but if we wish to understand the pain in relation to the body, we must learn to see how the psychological body schema can be aligned with the body, and how the pain can then be pointed to at a genuine location within the body and the wider physical world.

GLOSSARY

35–70 Hz hypothesis
The thesis, associated with Francis Crick and Christof Koch, that the "neural correlates" of consciousness are the firing of neurons at a frequency of 35–70 Hz.

Anomalous monism
The view, associated with Donald Davidson, according to which mental events are physical events and there are no strict laws governing the interaction of mental and physical events. The description of events as mental gives them an anomalous character; they do not obey laws. An analogy: "cheap" is an anomalous description, as there are no strict economic laws describing the cheapness of things (cf. "price"), yet every cheap thing is an economic object.

Attention schema
Michael Graziano's theory that consciousness is a simplified model or schema in our brains of our activities of attention or information filtering.

Behaviorism
The theory that descriptions of states of mind are or are a function of the tendency of the physical body to behave in specified ways.

Conservation law
Law of physics that the net total of mass or energy in a "closed" system stays the same. (There also exist conservation laws of other properties, such as linear momentum.)

Double aspect theory
The theory that mind and matter are two aspects of one neutral thing, just as a book could be viewed by a librarian under the aspect of price and under the aspect of subject matter, in Ryle's metaphor.

Eliminativism
The theory that mental categories and descriptions will not map onto the descriptions of a completed neuroscience, so mental descriptions will or should

be eliminated. The mentalistic terms of folk psychology ("want," "hope," "love," for example) do not describe or explain any physiological states, and so there are no such things as wanting, hoping and loving, just as the term "witch" describes nothing in a complete and true view of the world.

Emergentism
A view popular at the end of the nineteenth-century according to which mind *emerges* from matter when matter has reached a sufficient degree of complexity.

Epiphenomenalism
The theory that the mental phenomena are by-products of the basic physical phenomena, and that the epiphenomena do not in turn causally act on the physical phenomena.

Extension
Having length, or breadth or height, or x, y, or z coordinates in physical space.

Functionalism
The theory that mental states are functional states of organisms, and that functional states are computational states. A natural view for researchers in artificial intelligence.

Global workspace theory
A view developed by Bernard Baars according to which information in the brain converges on a particular kind of activity of a particular set of neurons, and the result of this convergence of information is identical with consciousness. A form of the identity theory applied to consciousness.

Idealism
The metaphysical view according to which everything is ideal, mental, or spiritual. Physicalism is a contrary view.

Identity theory
The view that the mind is identical with a set of brain processes.

Inconsistent tetrad
A group of four propositions, any three of which imply the falsity of the fourth.

Integrated information, theory of
The theory, advanced by Giorgio Tononi, that consciousness is integrated information in the brain.

Interactionism
The claim that the mind and body interact.

Mysterianism
The thesis that the mind–body problem is insoluble, and the mind is a mystery.

Neutral monism
The view of mind and body according to which things such as colors are neither physical nor mental, but "neutral" with respect to the mental and physical, except as they are placed by us into physical or a psychological explanatory relations. A manifestation of a color, according to Ernst Mach, is physical if we consider its relation to the luminous source, to temperature, and so on, and it is psychological if we consider its dependence on the retina, on the state of mind, and so on, in which case it is a sensation.

Panpsychism
The theory that everything in the universe has some level of mentality or consciousness.

Parallelism
A version of dualism that denies the interaction between mind and body.

Physical
Having a position in three-dimensional space; or having mass/energy; or being referred to in physics.

Physicalism
The view that the only things that exist are physical things. This view does not deny that mental things exist, provided that they are physical.

Property dualism
The theory that mental properties and physical properties are distinct, combined with the view that mind and body are not distinct substances.

Qualia

Allegedly indefinable properties of experience, or sometimes experiences themselves, such as "the redness of red," which are said to be "what it is like" to be conscious in that particular way.

Substance dualism

The theory that there are two fundamental types of individual things, mental and physical ones, and that they are distinct, in the sense that they can exist independently of one another.

Supervenience

A relationship between higher-level properties and lower-level properties. The higher-level properties H are said to supervene on the lower-level properties L if and only if there cannot be a change in H without a corresponding change in L. So aesthetic properties supervene on physical ones; a painting cannot improve aesthetically, for example, without a corresponding change in its physical composition. On supervenience views, mental properties are said to supervene on physical ones.

NOTES

Preface

1. David Chalmers, "Facing Up to the Problem of Consciousness," *Journal of Consciousness Studies* 2, no. 3 (1995): 200–219, and *The Conscious Mind* (Oxford: Oxford University Press, 1996).

2. Schopenhauer *did* use the word "der Weltknoten" ("the world-knot"), but he used it to describe the identity of the self in the unity of cognition and willing. The world-knot is the unity of the self not with the *body* but with the *world*, and the world-knot is not the mind–body problem at all. "The identity of the subject of willing with that of knowing by virtue whereof (and indeed necessarily) the word 'I' includes and indicates both, is the knot of the world, and hence inexplicable" (Arthur Schopenhauer, *On the Fourfold Root of the Principle of Sufficient Reason* [La Salle: Open Court, 1974], p. 210). See also Günther Zöller, "Schopenhauer on the Self," in *The Cambridge Companion to Schopenhauer*, ed. Christopher Janaway (Cambridge: Cambridge University Press, 1999), p. 26. For Schopenhauer, the world-knot cannot be explicated ("ist ... unerklärlich"), because we can only grasp the connection between *objects*, and the subject is not an object.

Chapter 1

1. See Keith Campbell, *Body and Mind* (New York: Anchor Books, 1984), 14. I have made the following changes in Campbell's formulation. For his "spiritual" I have substituted "nonphysical," for his "material" I have substituted "physical," and for his "do not interact" I have substituted "cannot interact." There are other formulations related to Campbell's that perhaps go deeper, but I have preferred to work with the simplest available formulation. For example, Kirk Ludwig gives the following formulation: (1) Some things have mental properties (realism); mental properties are not conceptually reducible to nonmental properties, and, consequently, no nonmental proposition entails any mental proposition (conceptual autonomy); (2) a complete description of a thing in terms of its basic constituents, their nonrelational properties, and (3) relations to one another and to other basic constituents of things, similarly described (the constituent description), entail a complete description of it, i.e., an account of all of a thing's properties follows from its constituent descriptions (constituent explanatory sufficiency); (4) the basic constituents of things do

not have mental properties as such (constituent nonmentalism). See Kirk Ludwig, "The Mind–Body Problem: An Overview," in *Blackwell Guide to Philosophy of Mind*, ed. Stephen Stich and Ted A. Warfield (Oxford: Blackwell, 2003), 10–11. The idea is that physical things, including the body, which does not feature in Ludwig's tetrad, have no mental properties, but mental properties do exist. Yet these properties cannot be explained by physical properties. Ludwig's tetrad is very interesting, though it does use the concepts of *constituent*, *reducible*, *as such*, the denial of *emergence*, and *relationality*. If we can avoid it, it seems to me that the mind–body problem has enough difficulties without introducing a list of complex and troubled concepts such as this one.

2. Timothy L. S. Sprigge, "Final Causes," *Aristotelian Society* suppl. vol. 45 (1971): 166ff.; Thomas Nagel, "What Is It Like to Be a Bat?" *Philosophical Review* 83, no. 4 (1974): 435–450.

3. Noam Chomsky, "Language and Nature," *Mind* 104, no. 413 (1994): 4–6. See also Ned Markosian, "What Are Physical Objects?" *Philosophy and Phenomenological Research* 61, no. 2 (2000): 375–395. Markosian defines physical objects as objects with spatial locations, Cartesian style, and defends his sensible view against the alternatives.

4. See Barbara Montero, "What Is the Physical?" in *The Oxford Handbook of the Philosophy of Mind*, ed. Brian McLaughlin, with Ansgar Beckermann and Sven Walter (Oxford: Oxford University Press, 2005), 173–188.

5. René Descartes, *The Philosophical Writings of Descartes*, vol. 2, ed. John Cottingham, Robert Stoothoff, and Dugald Murdoch (Cambridge: Cambridge University Press, 1984), 275.

6. Princess Elisabeth of Bohemia to Descartes, May 6–16, 1643, in *Descartes: Philosophical Writings*, trans. Elizabeth Anscombe and Peter Thomas Geach (London: Nelson, 1969), 274–275.

7. Gassendi, from "Fifth Set of Objections," in Descartes, *Philosophical Writings*, vol. 2, 234.

Chapter 2

1. E. J. Lowe, "The Problem of Psychophysical Causation," *Australasian Journal of Philosophy* 70, no. 3 (1992): 271–273.

2. Jaegwon Kim, "Lonely Souls: Causality and Substance Dualism," in K. Corcoran, ed., *Soul, Body, and Survival: Essays in the Metaphysics of Human Persons*, ed. K. Corcoran (Ithaca: Cornell University Press, 2001), 31.

3. A. J. Ayer, *The Physical Basis of Mind: A Sequence of Broadcast Talks*, ed. Peter Laslett (Oxford: Blackwell, 1951), 73–74. Ayer may have been influenced,

especially in his use of the metaphor of the two sets of *dramatis personae*, by George Stuart Fuller's discussion of the work of the parallelist W. K. Clifford given in Fuller's *A System of Metaphysics* (New York: Macmillan, 1904), 311: "the *dramatis personae* to whom we seem to be introduced at the outset are: an external object, which we will call a square, a retinal image of that object, which is also square; a disturbance of the ganglion, which we have no reason to believe square; and a mental image, which is a square."

4. Matthew Bennett, Michael F. Schatz, Heidi Rockwood, and Kurt Wiesenfeld, "Huygens's Clocks," *Proceedings of the Royal Society A*, 458 (2002): 563–579.

5. Ernest Lepore and Barry Loewer, "Mind Matters," *Journal of Philosophy* 84, no. 11 (1987): 630.

6. T. H. Huxley, "On the Hypothesis That Animals Are Automata, and Its History," *Fortnightly Review* 95 (1874): 240.

7. Stephen Law, "Honderich and the Curse of Epiphenomenalism," *Journal of Consciousness Studies* 13, nos. 7–8 (2006): 61–77.

Chapter 3

1. Gilbert Ryle, *The Concept of Mind*, 60th anniv. ed. (London: Routledge, 2009; first published, London: Hutchinson, 1949), 14.

2. Gilbert Ryle, *On Thinking*, ed. Konstantin Kolenda (Oxford: Blackwell, 1979), x.

3. Saul Kripke, *Naming and Necessity* (Cambridge, MA: Harvard University Press, 1980).

4. Central-state materialists used to write that pain is C-fiber stimulation, which is wrong because: (1) pain also involves the much faster myelenated $A\delta$-fiber transmissions to the spinal cord; (2) neither C- nor $A\delta$-fibers are part of the central nervous system at all; and (3) the peripheral $A\delta$- and C-fibers merely transmit information to the spinal cord. No one should identify the causes of pain with pain. The relevant states in the central nervous system are indeed within the thalamus, the prefrontal cortex, and the primary and secondary somatosensory cortex (S1 and S2). See Roland Puccetti, "The Great C-Fiber Myth: A Critical Note," *Philosophy of Science* 2, no. 44 (1977): 303–305.

5. Hilary Putnam, "Psychological Predicates," in *Art, Mind, and Religion*, ed. W. H. and D. D. Merrill (Pittsburgh: University of Pittsburgh Press, 1967), reprinted as "The Nature of Mental States," in *Mind, Language and Reality: Philosophical Papers*, vol. 2 (Cambridge: Cambridge University Press, 1975).

6. Donald Davidson, "Mental Events," in *Essays on Actions and Events* (Oxford: Oxford University Press, 1980), 207–227.

7. Paul Churchland, "Eliminative Materialism and the Propositional Attitudes," *Journal of Philosophy* 2, no. 78 (1981): 67.

8. Churchland, "Eliminative Materialism," 76.

Chapter 4

1. Jaegwon Kim, *Philosophy of Mind* (Boulder, CO: Westview, 2006), 44–48; Jaegwon Kim, *Physicalism, or Something Near Enough* (Princeton, NJ: Princeton University Press, 2005).

2. Others who have taken an antimaterialist line, in very well-known works, are John Searle, Joseph Levine, and Colin McGinn.

3. Nagel, Thomas, "What Is It Like to Be a Bat," *Philosophical Review* 83, no. 4 (1974): 439, reprinted in his *Mortal Questions* (Cambridge: Cambridge University Press, 2000). "I am what is sometimes known as a 'qualia freak,'" Frank Jackson writes, in a similar vein:

I think that there are certain features of bodily sensations especially, but also of certain perceptual experiences, which no amount of purely physical information includes. Tell me everything physical there is to tell about what is going on in a living brain ... you won't have told me about the hurtfulness of pains, the itchiness of itches, pangs of jealousy, or about the characteristic experience of tasting a lemon, smelling a rose, hearing a loud noise or seeing the sky. (Frank Jackson, "Epiphenomenal Qualia," *Philosophical Quarterly* 32, no. 127 [1982]: 127)

4. David Chalmers, "Facing Up to the Problem of Consciousness," *Journal of Consciousness Studies* 2 (1995): 3; David Chalmers, *The Conscious Mind* (Oxford: Oxford University Press, 1996).

Chapter 5

1. Bernard Baars, "Global Workspace Theory of Consciousness: Toward a Cognitive Neuroscience of Human Experience?" *Progress in Brain Research* 150 (2005), pp. 47–48.

2. Scott O. Murray, Daniel Kersten, Bruno A. Olshausen, Paul Schrater, and David L. Woods, "Shape Perception Reduces Activity in Human Primary Visual Cortex," *Proceedings of the National Academy of Sciences* 99, no. 23 (2002): 15164–15169.

3. Francis Crick and Christof Koch, "Towards a Neurobiological Theory of Consciousness," *Seminars in the Neurosciences* 2 (1990): 263.

4. Francis Crick, *The Astonishing Hypothesis: The Scientific Search for the Soul* (New York: Scribner's, 1994), 7.

5. Francis Crick and Christof Koch, "A Framework for Consciousness," *Nature Neuroscience* 6, no. 2 (2003): 123.

6. Francis Crick and Christof Koch, "What Is the Function of the Claustrum?" *Philosophical Transaction of the Royal Society*, 360, no. 1458 (2005): 1276.

7. Giulio Tononi, "Consciousness as Integrated Information: A Provisional Manifesto," *Biological Bulletin* 215, no. 3 (2008): 216–242.

8. See Michael S. A. Graziano, *Consciousness and the Social Brain* (Oxford: Oxford University Press, 2013), and Taylor W. Webb and Michael S. A. Graziano, "The Attention Schema Theory: A Mechanistic Account of Subjective Awareness," *Frontiers in Psychology* 6 (2015), article 500.

Chapter 6

1. John Locke, *An Essay Concerning Human Understanding*, ed. Pauline Phemister (Oxford: Oxford University Press, 2008), Book II, viii, 9–10.

2. Ludwig Wittgenstein, *Philosophical Investigations* (Oxford: Blackwell, 2002), 109e–110e, para. 308.

3. Gilbert Ryle, *The Concept of Mind* (London: Hutchinson, 1949), 17.

4. Ryle, *The Concept of Mind*, 23.

Chapter 7

1. Ernst Mach, *The Analysis of Sensations*, trans. C. M. Williams (New York: Dover, 1959), 29.

2. See Eric C. Banks's *The Realistic Empiricism of Mach, James, and Russell* (Cambridge: Cambridge University Press, 2014).

3. Banks, *Mach, James, Russell*, 6.

4. Banks, *Mach, James, Russell*, 6, 20.

5. Mach, *Analysis of Sensations*, 2.

6. Banks, *Mach, James, Russell*, 20.

7. For more on Russellian monism, and some interesting material on genuinely neutral monism, see Torin Alter and Yujin Nagasawa, eds., *Consciousness in the Physical World: Perspectives on Russellian Monism* (Oxford: Oxford University Press, 2015), especially on panqualityism (which Chalmers discusses in his chapter of Alter and Nagasawa's volume, "Panpsychism and Panprotopsychism," 270ff.), a view having something in common with neutral monism and with Russell's idea of *sensibilia* or unsensed sense data: the ultimate constituents of the world are neutral qualitative elements, such as colors, which can exist even when they are not perceived. What panqualityism lacks is the idea that these constituents can be placed in the functional context of either a

physical or a psychological sequence. Worse, for panqualityism, the fundamental qualities can attach to fundamental particles, so there is no knowing how they relate to what we see or hear or otherwise experience.

8. Mach, *Analysis of Sensations*, 29–30.

9. See Donovan Wishon, "Russell on Russellian Monism," in *Consciousness in the Physical World*, 104, and Banks, *Mach, James, Russell*, 16. See also Jonathan Westphal, "'Ernst Mach': An Immediate Joy in Seeing," *Times Literary Supplement* 4907 (1997): 7–8.

10. Ernst Mach, *The Analysis of Sensations*, 27.

11. Mach, *The Analysis of Sensations*, 17.

12. Mach, *The Analysis of Sensations*, 17–18.

13. Bertrand Russell, *Theory of Knowledge: The 1913 Manuscript* (London: Allen & Unwin, 1984), 15. This work is critical of neutral monism, and in the passage cited Russell is merely stating the theory; but by 1919 he had made it his own view, in "On Propositions."

14. Moritz Schlick, "On the Relation between Psychological and Physical Concepts," trans. Wilfrid Sellars, in *Philosophical Papers*, vol. 2 (1925–1936), ed. H. Mulder and Barbara F. B. van de Velde-Schlick (Dordrecht: D. Reidel, 1979), 431.

15. David Chalmers, "Panpsychism and Panprotopsychism," in *Consciousness in the Physical World*, 270–271.

16. S. Bourane, Martyn Goulding et al., "Gate Control of Mechanical Itch by a Subpopulation of Spinal Cord Interneurons," *Science* 35, no. 6260 (2015): 550–554.

17. G. W. Leibniz, *New Essays on Human Understanding*, ed. and trans. Peter Remnant and Jonathan Bennett (Cambridge: Cambridge University Press, 1981), 137.

18. Bertrand Russell, "On Propositions: What They Are and How They Mean," *Proceedings of the Aristotelian Society*, suppl. vol. 2 (1919), reprinted in Robert Charles Marsh, Bertrand Russell, *Logic and Knowledge* (London: Allen & Unwin, 1956), 299.

19. Here I develop further a model of the inconsistent tetrad that I originally proposed in *Philosophical Propositions* (London: Routledge, 1998), 128–130. I hope that this model is in the spirit of what Timothy Williamson argues for in "Model-Building in Philosophy," to appear in *Philosophy's Future: The Problem of Philosophical Progress*, ed. Russell Blackford and Damien Broderick (Oxford: Wiley, forthcoming).

20. Indigo is not a "familiar" color, if it is a color at all, or if it even exists. At the least, for most people the seven spectral colors described by Newton

(ROYGBIV) do not all have the same or a uniform status as colors; indigo, and perhaps orange as well, seem to be distinctly "secondary" compared with red, yellow, green, blue, and violet. For Newton's sophisticated mathematical shoe-horning ("The analogy of nature is to be preserved") of the colors into the seven intervals of the musical scale, see Peter Pesic, "Isaac Newton and the Mystery of the Major Sixth: A Transcription of His Manuscript 'Of Musick' with Commentary," *Interdisciplinary Science Reviews* 31, no. 4 (2006): 291–306.

21. It is not hard to find many more tetrads that work in exactly the same way. Here is an example. Imagine a brick-walled city square that is by a perfectly extraordinary accident possessed of corners whose angles are each exactly 90°. (The "s" in "(1s)" is for "square.") (1s) A square is an abstract thing. (2s) The bulldozer is a concrete thing. (3s) The square and the bulldozer interact, when the bulldozer demolishes one wall of the square. (4s) Abstract things and concrete things cannot interact. One way of looking at the problem is to note the ambiguity of "square," as a between a concrete and an abstract entity. In no sequence of abstract things is the next term a concrete thing, and in no sequence of concrete things is the next term an abstract thing.

BIBLIOGRAPHY

Alter, T., and Y. Nagasawa, eds. *Consciousness in the Physical World: Perspectives on Russellian Monism*. Oxford: Oxford University Press, 2015.

Ayer, A. J. *The Physical Basis of Mind: A Series of Broadcast Talks*. Ed. Peter Laslett. Oxford: Blackwell, 1951.

Baars, Bernard. "Global Workspace Theory of Consciousness: Toward a Cognitive Neuroscience of Human Experience?" *Progress in Brain Research* 150 (2005): 45–53.

Banks, Eric C. *The Realistic Empiricism of Mach, James, and Russell*. Cambridge: Cambridge University Press, 2014.

Bennett, Matthew, Michael F. Schatz, Heidi Rockwood, and Kurt Wiesenfeld. "Huygens's Clocks." *Proceedings of the Royal Society of London, Series A* 458 (2002): 563–579.

Bohemia, Princess Elisabeth of. Letter to Descartes, May 6–16, 1643. In *Descartes: Philosophical Writings*, trans. Elizabeth Anscombe and Peter Thomas Geach. London: Nelson, 1969.

Bourane, S., M. Goulding, B. Duan, S. C. Koch, A. Dalet, O. Britz, L. Garcia-Campmany, E. Kim, L. Cheng, A. Ghosh, and Q. Ma. "Gate Control of Mechanical Itch by a Subpopulation of Spinal Cord Interneurons." *Science* 35, no. 6260 (2015): 550–554.

Campbell, Keith. *Body and Mind*. New York: Anchor Books, 1984.

Chalmers, David. "Facing Up to the Problem of Consciousness." *Journal of Consciousness Studies* 2, no. 3 (1995): 3.

Chalmers, David. *The Conscious Mind*. Oxford: Oxford University Press, 1996.

Chomsky, Noam. "Language and Nature." *Mind* 104, no. 413 (1994): 4–6.

Churchland, Paul. "Eliminative Materialism and the Propositional Attitudes." *Journal of Philosophy* 78, no. 2 (1981): 67.

Crick, Francis. *The Astonishing Hypothesis*. Oxford: Oxford University Press, 1995.

Crick, Francis, and Christof Koch. "A Framework for Consciousness." *Nature Neuroscience* 6, no. 2 (2003): 119–126.

Crick, Francis, and Christof Koch. "Towards a Neurobiological Theory of Consciousness." *Seminars in Neuroscience* 2 (1990): 263.

Crick, Francis, and Christof Koch. "What Is the Function of the Claustrum?" *Philosophical Transaction of the Royal Society* 360, no. 1458 (2005).

Davidson, Donald. "Mental Events." In *Essays on Actions and Events*. Oxford: Oxford University Press, 1980.

Descartes, René. *The Philosophical Writings of Descartes*, vol. 2. Ed. J. Cottingham, R. Stoothoff, and D. Murdoch. Cambridge: Cambridge University Press, 1984.

Fuller, George Stuart. *A System of Metaphysics*. New York: Macmillan, 1904.

Graziano, M. S. A. *Consciousness and the Social Brain*. Oxford: Oxford University Press, 2013.

Huxley, T. H. "On the Hypothesis That Animals Are Automata, and Its History." *Fortnightly Review* 95 (1874): 240.

Jackson, Frank. "Epiphenomenal Qualia." *Philosophical Quarterly* 32, no. 127 (1982): 127–136.

Kim, Jaegwon. "Lonely Souls: Causality and Substance Dualism." In *Soul, Body and Survival: Essays in the Metaphysics of Human Persons*, ed. K. Corcoran. Ithaca: Cornell University Press, 2001.

Kim, Jaegwon. *Philosophy of Mind*. Boulder, CO: Westview, 2006.

Kim, Jaegwon. *Physicalism, or Something Near Enough*. Princeton, NJ: Princeton University Press, 2005.

Kripke, Saul. *Naming and Necessity*. Cambridge, MA: Harvard University Press, 1980.

Law, Stephen. "Honderich and the Curse of Epiphenomenalism." *Journal of Consciousness Studies* 13, nos. 7–8 (2006): 61–77.

Leibniz, G. W. *New Essays on Human Understanding*. Ed. and trans. P. Remnant and J. Bennett. Cambridge: Cambridge University Press, 1981.

Lepore, Ernie, and Barry Loewer. "Mind Matters." *Journal of Philosophy* 84, no. 11 (1987): 630.

Locke, John. *An Essay Concerning Human Understanding*. Ed. Pauline Phemister. Oxford: Oxford University Press, 2008. (First published 1689.)

Lowe, E. J. "The Problem of Psychophysical Causation." *Australasian Journal of Philosophy* 70, no. 3 (1992): 3.

Ludwig, Kirk. "The Mind–Body Problem: An Overview." In *The Blackwell Guide to Philosophy of Mind*, ed. Stephen Stich and Ted A. Warfield. Oxford: Blackwell, 2003.

Mach, Ernst. *The Analysis of Sensations*. Trans. C. M. Williams. New York: Dover, 1959.

Markosian, Ned. "What Are Physical Objects?" *Philosophy and Phenomenological Research* 61, no. 2 (2000): 375–395.

Montero, Barbara. "What Is the Physical?" In *The Oxford Handbook of the Philosophy of Mind*, ed. Brian McLaughlin, with Ansgar Beckermann and Sven Walter. Oxford: Oxford University Press, 2005.

Murray, Scott O., Daniel Kersten, Bruno A. Olshausen, Paul Schrater, and David L. Woods. "Shape Perception Reduces Activity in Human Primary Visual Cortex." *Proceedings of the National Academy of Sciences of the United States of America* 99, 23 (2002): 15164–15169.

Nagel, Thomas. "What Is It Like to Be a Bat?" *Philosophical Review* 83, no. 4 (1974): 4. Reprinted in his *Mortal Questions* (Cambridge: Cambridge University Press, 2000).

Pesic, Peter. "Isaac Newton and the Mystery of the Major Sixth: A Transcription of His Manuscript 'Of Musick' with Commentary." Interdisciplinary Science Reviews 31, no. 4 (2006): 291–306.

Puccetti, Roland. "The Great C-Fiber Myth: A Critical Note." *Philosophy of Science* 2, no. 44 (1977): 303–305.

Putnam, Hilary. "Psychological Predicates." In Art, *Mind and Religion*, ed. W. H. Merrill and D. D. Merrill. Pittsburgh: University of Pittsburgh Press, 1967. Reprinted as "The Nature of Mental States," in *Mind, Language and Reality: Philosophical Papers*, vol. 2 (Cambridge: Cambridge University Press, 1975).

Russell, Bertrand. "On Propositions: What They Are and How They Mean." *Proceedings of the Aristotelian Society*, suppl. vol. 2 (1919). Reprinted in Bertrand Russell, *Logic and Knowledge* (London: Allen & Unwin, 1956).

Russell, Bertrand. *Theory of Knowledge: The 1913 Manuscript*. London: Allen & Unwin, 1984.

Ryle, Gilbert. *On Thinking*, ed. K. Kolenda. Oxford: Blackwell, 1979.

Ryle, Gilbert. *The Concept of Mind*, 60th anniversary ed. London: Routledge, 2009. (First published London: Hutchinson, 1949.)

Schlick, Moritz. "On the Relation between Psychological and Physical Concepts." Trans. Wilfrid Sellars. In *Philosophical Papers*, vol. 2 (1925–1936), ed. H. Mulder and Barbara F. B. van de Velde-Schlick. Dordrecht: D. Reidel, 1979.

Schopenhauer, Arthur. *On the Fourfold Root of the Principle of Sufficient Reason*. La Salle: Open Court, 1974.

Sprigge, Timothy L. S. "Final Causes." *Proceedings of the Aristotelian Society* suppl. vol. 45 (1971).

Tononi, Giulio. "Consciousness as Integrated Information: A Provisional Manifesto." *Biological Bulletin* 215, no. 3 (2008): 216–242.

Webb, Taylor W., and Michael A. Graziano. "The Attention Schema Theory: A Mechanistic Account of Subjective Awareness." *Frontiers in Psychology* 6 (2015), article 500.

Westphal, Jonathan. "'Ernst Mach': An Immediate Joy in Seeing." *Times Literary Supplement* 4907 (1997): 7–8.

Westphal, Jonathan. *Philosophical Propositions*. London: Routledge, 1998.

Williamson, Timothy. "Model-Building in Philosophy." In *Philosophy's Future: The Problem of Philosophical Progress*, ed. Russell Blackford and Damien Broderick. Oxford: Wiley, forthcoming.

Wishon, Donovan. "Russell on Russellian Monism." In *Consciousness in the Physical World: Perspectives on Russellian Monism*, ed. Torin Alter and Yujin Nagasawa. Oxford: Oxford University Press, 2015.

Wittgenstein, Ludwig. *Philosophical Investigations*. Oxford: Blackwell, 2002. (First published 1953.)

Zöller, Günther. "Schopenhauer on the Self." In *The Cambridge Companion to Schopenhauer*, ed. Christopher Janaway. Cambridge: Cambridge University Press, 1999.

FURTHER READINGS

Almog, J. *What Am I? Descartes and the Mind–Body Problem*. Oxford: Oxford University Press, 2001.

Armstrong, David. *The Mind–Body Problem: An Opinionated Introduction*. Boulder, CO: Westview, 1999.

Bricke, J. "Interaction and Physiology." *Mind* 84, no. 334 (1975): 255–259.

Clark, Austen. "Color, Qualia, and Attention: A Non-Standard Interpretation." In *Color Ontology and Color Science*, ed. Jonathan D. Cohen and Mohan Matthen. Cambridge, MA: MIT Press, 2010.

Crane, Tim, and Sarah Patterson. *History of the Mind–Body Problem*. London: Routledge, 2000.

Dennett, Daniel C. *Consciousness Explained*. Boston: Little, Brown, 1991.

Foster, J. "A Defence of Dualism." In *The Case for Dualism*, ed. J. Smythies and J. Beloff. Charlottesville, NC: University of Virginia Press, 1989.

Foster, J. "Psycho-Physical Causal Relations." *American Philosophical Quarterly* 5, no. 1 (1968): 64–70.

Himma, K. E. "What Is a Problem for All Is a Problem for None: Substance Dualism, Physicalism, and the Mind–Body Problem." *American Philosophical Quarterly* 42, no. 2 (2005): 81–92.

Jaworski, William. *Philosophy of Mind*. Oxford: Oxford University Press, 2011.

Lycan, William. "Chomsky on the Mind–Body Problem." In *Chomsky and His Critics*, ed. Louise Anthony. Malden, MA: Blackwell, 2003.

McGinn, C. "Can We Solve the Mind–Body Problem?" *Mind* 98, no. 398 (1989): 349–356.

Nagel, Thomas. *The View from Nowhere*. Oxford: Oxford University Press, 1986.

Penrose, Roger, and Stuart Hameroff. "Consciousness in the Universe: Neuroscience, Quantum Space-Time Geometry, and Orch OR Theory." *Journal of Cosmology* 14 (2011).

Rorty, Richard. *Philosophy and the Mirror of Nature*. Princeton: Princeton University Press, 1980.

Searle, John. *The Mystery of Consciousness*. New York: New York Review of Books, 1997.

Searle, John. *The Rediscovery of the Mind*. Cambridge, MA: MIT Press, 1992.

Taylor, Richard. "How to Bury the Mind–Body Problem." *American Philosophical Quarterly* 6, no. 2 (1969): 136–143.

Tye, Michael. *Consciousness Revisited*. Cambridge, MA: MIT Press, 2009.

Vision, Gerald. *Re-Emergence: Locating Conscious Properties in a Material World*. Cambridge, MA: MIT Press, 2011.

INDEX

JONATHAN WESTPHAL is a Permanent Member of the Senior Common Room at University College, Oxford. He is the author of *Colour: A Philosophical Introduction*.